Magnetism
and Transition
Metal Complexes

Magnetism and Transition Metal Complexes

F. E. Mabbs and D. J. Machin

Lecturers in Chemistry
at the University of Manchester

London
Chapman and Hall

First published 1973
by Chapman and Hall Ltd
11 New Fetter Lane, London EC4P 4EE
© 1973 F.E. Mabbs and D.J. Machin
Printed in Great Britain by
William Clowes & Sons Limited
London, Colchester and Beccles

SBN 412 11230 2

Distributed in the U.S.A.
by Halsted Press, a Division
of John Wiley & Sons, Inc.
New York

Contents

Preface

The use of the magnetic properties of d-transition metal complexes as a means of determining stereochemistry and ground state electronic properties has increased very greatly since the late 1950's. In the course of this growth, several crises of confidence have been faced. Thus it was initially assumed that the room temperature magnetic moment would yield both the oxidation state of the metal and the stereochemistry of the complex, based on the observation of deviations from the 'spin-only' magnetic moment. The many anomalies which this procedure revealed led to the use of measurements on powders over a temperature range (usually 80–300 K), with deductions based on the temperature dependence of the magnetic moment. This also proved to be a somewhat risky procedure with the recognition that very few complexes had the strict O_h or T_d symmetry postulated in the theory, and also with the increasing realization that these were perhaps not even the most common stereochemistries. Effects due to covalency were also considered at this stage.

The situation now is that in order to allow for these refinements it is necessary to introduce more parameters than can be reliably determined from measurements of average susceptibilities down to 80 K. This does not mean that the method is discredited, but rather that more extensive measurements and more careful interpretation are needed in order to obtain the information required. Thus a study of the principal susceptibilities of single crystals over a wider range of temperature, perhaps together with electron spin resonance parameters can still yield very valuable information.

With the advent of good commercial apparatus, the study of magnetic properties is available to all inorganic chemists. However there is an almost complete lack of books setting out in any detail the theoretical basis which must be understood before the chemist can *reliably* interpret his data, or set out to do his own calculations. We seek to fill this gap. The theory on which the calculation of magnetic properties is based is not difficult, and is certainly not beyond the grasp of any chemist. This book is based upon lecture courses which we give to final-year undergraduate and first-year postgraduate students in the University of Manchester: our experience is that they can cope with the theory at this level provided that its application is widely illustrated. We have not sought to be rigorous in our treatment of the quantum mechanical basis of the calculations. Many good texts for the expert in this area already exist. Instead we have sought to illustrate the *results* of the theory and show in detail how they are applied in actual systems. It is impossible to deal with all known stereochemistries: the reader who understands the treatment of cubic (i.e. O_h or T_d) complexes and axially distorted cubic complexes which we discuss should be able to extend the calculations to other systems in which he is interested. We have also restricted the discussion to complexes containing unpaired d-electrons.

The book is set out as follows: Chapter 1 contains the formal definitions of magnetic quantities and types of magnetic behaviour. This is followed by a general, qualitative discussion of the magnitudes and temperature dependence of magnetic moments.

Chapter 2 is a brief survey of crystal field theory, setting out the results which will be needed in subsequent chapters. Chapter 3 deals with perturbation theory mainly in an illustrative manner. This is followed by some general description of the application of perturbation theory to spin-orbit coupling and magnetic field perturbations.

Chapters 4 and 5 contain detailed calculations of the magnetic properties of 'cubic' and axially distorted cubic complexes respectively, as well as general discussions of the results obtained.

Chapter 6 is devoted to a discussion of a small number of magnetic studies taken from the literature. The examples have been chosen so as to illustrate both strengths and weaknesses in the interpretations and in the data, and ways in which more extensive studies might have improved the interpretation. We do not provide any comprehensive literature survey.

Chapter 7 contains a discussion of polynuclear complexes. Here we are concerned with magnetic interactions between small numbers (rarely more than three) of paramagnetic centres in a single molecule. Such interactions are presumed to be absent in the discussion in preceding chapters. There has been considerable interest in the magnetic properties of these polynuclear complexes in recent years. We describe a theoretical approach to the problem and discuss a number of examples. We do not consider lattice interactions, i.e. those in which the magnetic interactions extend over large numbers of centres.

Each chapter is completed by a list of references to both books and journals. These have been chosen to illustrate particular points, and are in no way comprehensive.

Finally, it is a pleasure to record our gratitude to Professor J. Lewis and the late Sir Ronald Nyholm, who introduced us both to magnetochemistry. We would also like to thank them, as well as Dr B.N. Figgis and Dr M. Gerloch for many stimulating discussions of magnetochemistry since then.

Manchester, June 1972.   F.E. Mabbs

D.J. Machin

Symbols and useful constants

Symbols used in the text

$A, B, C*$	(see comment below)
a_0	First Bohr radius
$c*$	Speed of light (see comment below)
Cp	Axial crystal field parameter
Dq	Cubic crystal field parameter
e	Electronic charge
F_i	Condon-Shortley interelectron repulsion parameters (i = 0,2,4 for d-electrons)
g	Gyromagnetic ratio for an electron
H	Magnetic field
$\mathscr{H}$	Hamiltonian operator
h	Planck's constant
$\hbar$	$h/2\pi$
I	Intensity of magnetization
I_{mn}	A general matrix element $\langle \psi_m \mid \text{operator} \mid \psi_n \rangle$
i	$(-1)^{\frac{1}{2}}$
k	Boltzmann's constant
$\mathbf{k}$	Orbital reduction factor
N	Avogadro's constant
$N\alpha$	Temperature independent magnetic susceptibility
$R(r)$	Radial part of an electronic wave-function
T	Absolute temperature
T_C	Curie temperature

T_N	Néel temperature
V_{oct}	Octahedral crystal field potential
V_{tetrag}	Tetragonal crystal field potential
V_{trig}	Trigonal crystal field potential
v	Δ/λ
$Y_1^{m1}(\theta,\phi)$ or Y_1^{m1}	Angular part of an electronic wave-function (spherical harmonic)
α	A general parameter (not to be confused with Nα given above)
β	Bohr magneton (B.M.)
Δ	Separation between two energy states
δ	Zero field splitting
ϵ	Constitutive diamagnetic correction
$\zeta_{n,1}$	Single electron spin-orbit coupling constant
η	$B_{complex}/B_{free\ ion}$ (B is the Racah parameter)
θ	Weiss constant
κ	Magnetic susceptibility per unit volume
λ	Spin-orbit coupling constant for a term
μ	Magnetic moment (effective magnetic moment)
$\mu_{s.o.}$	Spin only magnetic moment
Φ,ϕ Ψ,ψ	Wave-functions
χ_g	Magnetic susceptibility per gram
χ_{mol}	Magnetic susceptibility per molecule
χ_M	Magnetic susceptibility per mole
χ_A	Magnetic susceptibility per mole of atoms
$\chi_{\parallel}$	Magnetic susceptibility per mole of atoms parallel to the principal axis
$\chi_{\perp}$	Magnetic susceptibility per mole of atoms perpendicular to the principal axis

Operators are denoted by a circumflex. Thus 1 is a quantum number, but $\hat{1}$ is an orbital angular momentum operator. The conventional symbols are always used for quantum numbers.

* Unfortunately, certain symbols are commonly used to represent a variety of variables. The following are used with several meanings in the text:

A A Racah interelectron repulsion parameter (used only in Section 2.2.1), also a parameter representing the mixing of certain wave-functions in a crystal field (introduced in Section 2.3.3 and used subsequently in many places)

B Magnetic induction (used only in Chapter 1), also a Racah interelectron repulsion parameter

C A Racah interelectron repulsion parameter, also the Curie constant in the Curie law (i.e. $\chi = C/T$)

c Speed of light. This symbol is also commonly used as a mixing coefficient in wavefunctions. In the latter connection it always carries a subscript.

Useful constants

Boltzmann's constant, k, $1 \cdot 38044 \times 10^{-16}$ erg deg^{-1} mole^{-1}

Planck's constant, h, $6 \cdot 6256 \times 10^{-27}$ erg s

Bohr magneton, β, $\begin{cases} 0 \cdot 92731 \times 10^{-20} \text{ erg gauss}^{-1}, \text{ or} \\ 4 \cdot 66858 \times 10^{-5} \text{ cm}^{-1} \text{ gauss}^{-1} \end{cases}$

Avogadro's constant, N, $6 \cdot 0249 \times 10^{23}$ atoms mole^{-1}

g (free electron) $2 \cdot 0023$

velocity of light, c, $2 \cdot 9979 \times 10^{10}$ cm s^{-1}

$N\beta^2$ $0 \cdot 26073$ cm^{-1} erg gauss^{-2} mole^{-1}*

$3k/N\beta^2$ $7 \cdot 9971$ mole gauss^2 erg^{-1} deg^{-1}

kT $1 \cdot 3804 \times 0 \cdot 5035T$ cm^{-1} mole^{-1}

    (e.g. $208 \cdot 4$ cm^{-1} mole^{-1} at 300 K)

---

* These dimensions are given because this is the most convenient form when used in magnetic calculations.

Units in magnetism

This book is written using the unrationalized c.g.s. system of electromagnetic units which has been the standard practice of magnetochemists for many years. SI units are rationalized, and so the conversion from c.g.s. e.m.u. to SI magnetic units involves factors which include 4π, since the permeability of free space is $4\pi \times 10^{-7}$ kg m $s^{-2}A^{-2}$ in SI, but is one, and dimensionless, in the c.g.s. system. Thus there is not always a simple factor of ten conversion between the two systems. For this reason we have retained the well-known c.g.s. system.

The relevant conversions to SI units have been discussed by a number of authors recently. We summarize the most important of these below, and refer the interested reader to the literature cited below.

Quantity	Symbol		SI unit
	c.g.s. e.m.u.	SI	c.g.s. unit
Magnetic field	oersted	Am^{-1}	$10^3/4\pi$
Magnetic induction	gauss	$\begin{cases} Wb\ m^{-2} \\ (\text{Tesla}) \end{cases}$	10^{-4}
Volume susceptibility	dimensionless	dimensionless	4π
Gram susceptibility	$cm^3\ g^{-1}$	$m^3\ kg^{-1}$	$4\pi/10^3$
Molar susceptibility	$cm^3\ mole^{-1}$	$m^3\ mol^{-1}$	$4\pi/10^6$
Magnetic moment	Bohr magneton	Bohr magneton	1*

* In the c.g.s. system the Bohr magneton is $0\cdot92731 \times 10^{-20}$ erg gauss^{-1} and in SI it is $0\cdot92731 \times 10^{-23}$ A m^2 molecule^{-1}. Using these values, the magnetic moment, μ, is $2\cdot8279$ (χ_A T)$^{\frac{1}{2}}$ Bohr magnetons in c.g.s. units and $7\cdot9774 \times 10^2$ (χ_A T)$^{\frac{1}{2}}$ Bohr magnetons in SI units. These two expressions give the same *numerical* value for μ, so that expressions such as $\mu = [4S(S + 1)]^{\frac{1}{2}}$ Bohr magnetons are valid in either system of units.

REFERENCES

1. G.H. Rayner and A.E. Drake (1970). *SI Units in Electricity and Magnetism*, HMSO, London.
2. G. Pass and H. Sutcliffe (1970). *J. Chem. Education*, 180.
3. T.I. Quickenden and R.C. Marshall (1972). *J. Chem. Education*, 114.

(The last reference includes a general description of rationalized and un-rationalized units, and is recommended as a general account of magnetic units.)

1 General considerations

1.1. Introduction

The past fifteen years have seen a considerable increase in the use, by chemists, of magnetic properties as a tool for the determination of the structure of d-transition metal complexes. It is a sobering thought that the theories required to calculate these properties have been known for some forty years. This is particularly so when it is realized that many chemists still do not understand the theoretical basis of their structural assignments. A lack of appreciation of the limitations as well as the uses of magnetic studies is common: this book sets out to provide the chemist with a knowledge of the theory, and by the use of a small number of examples to illustrate both strengths and weaknesses of the approach.

The subject matter is developed first by introducing the basic terminology and making some generalizations about magnetic behaviour. The second chapter contains a brief survey of the energy levels of free ions, of crystal field theory and of spin-orbit coupling. We shall use an approach based upon the crystal field theory. This is not to say that a ligand field (i.e. molecular orbital) approach cannot be used; rather that the basically simpler point-charge approach yields virtually all of the answers which we require. The principal method for calculating the effects of crystal fields, etc. is the perturbation theory. This is the subject of Chapter 3. It is common at this stage for chemists to say 'Oh, I don't understand this mathematical stuff': perturbation theory is superficially complex, mainly because of complexities of nomenclature rather than of mathematics. In Chapter 3 we develop the theory in an illustrative manner; some effort spent here will greatly increase the enjoyment and value of the succeeding chapters.

Chapters 4 and 5 deal with the magnetic properties of d-transition metal complexes; we do not consider complexes with unpaired f-electrons in this book. Chapter 4 deals with the magnetism of complexes with strict O_h or T_d symmetry (it is common to describe these complexes as 'cubic' and we shall do so. Strictly O_h and T_d are sub-groups of the cubic group). Chapter 5 considers the effects of an axial distortion along a four-fold or a three-fold axis. In Chapter 6 we discuss a number of examples taken from the literature. These are chosen to illustrate both the strengths and weaknesses of magnetic studies. We make no attempt to survey published data comprehensively.

Chapter 7 is devoted to a discussion of the magnetic properties of polynuclear complexes. In all of the preceding discussion it is assumed that the magnetic behaviour of a particular atom is independent of that of its neighbours. This presumption is often unjustified in polynuclear species and there has been considerable interest in such molecules which have only a small number of interacting centres. An understanding of the resulting antiferromagnetism (only occasionally ferromagnetism) is the subject of Chapter 7. We do not discuss magnetic interactions in extended lattices.

The study of magnetic properties can reveal much about the detailed electronic structure as well as information about the stereochemistry of transition metal complexes. However, this information is only acquired by a careful magnetic study, usually on single crystals, over a wide temperature range. It is essential to appreciate the *reliability* of the interpretation given to them.

We shall not deal explicitly with the closely related phenomenon of electron spin resonance although much of the calculation is appropriate thereto. Neither shall we discuss experimental methods since these are well documented (see for example references [1–3]).

1.2. Definitions

We shall use unrationalized c.g.s. electromagnetic units throughout this book. The SI units are, of course, a rationalized system of units, and so are related to the more familiar c.g.s. units which we use by factors which include 4π. There is thus no simple power of ten relationship between the two systems of units. For this reason SI units have not been readily accepted by practising magneto-chemists. For those who wish to use SI units, some further comments, and the necessary conversion factors will be found on page xvii of the introduction.

If a body is placed in an homogeneous magnetic field, H, the field within the body generally differs from the free space value. The field within the body, the *magnetic induction B*, is then

$$B = H + \Delta H.$$

The body has, in effect, become magnetized, and ΔH is readily related to the *intensity of magnetization I*, of the body. Thus

$$B = H + 4\pi I. \tag{1.1}$$

A more useful quantity than I is the *volume magnetic susceptibility*, κ, which is related to I by

$$\kappa = I/H. \tag{1.2}$$

Dividing Equation 1.1 by H gives

$$B/H = 1 + 4\pi\kappa. \tag{1.3}$$

Thus any experimental method for determining magnetic susceptibilities depends ultimately upon the measurement of B/H. It is readily shown that an element of volume dv in a magnetic field H and gradient $\partial H/\partial x$ will experience a force df where

$$df = (\kappa - \kappa_0)dvH\,\partial H/\partial x \tag{1.4}$$

where κ and κ_0 are the volume susceptibilities of the body and its surroundings respectively. The two principal experimental methods for measuring κ treat Equation 1.4 in two ways. In the Faraday method it is arranged that $H\partial H/\partial x$ is constant over the whole body whence

$$F = (\kappa - \kappa_0)VH\,\partial H/\partial x.$$

The Guoy method arranges the sample so that the field is a maximum (H) at one end and zero or constant (H_0) at the other when integration of Equation 1.4 gives

$$F = \tfrac{1}{2}(\kappa - \kappa_0)V(H^2 - H_0^2).$$

It is generally more convenient to measure mass rather than volume, thus the *magnetic susceptibility per gram* of material, χ_g, is

$$\chi_g = \kappa/\rho$$

where ρ is the density of the material.

Similarly, the *susceptibility per mole* is

$$\chi_M = \chi_g M$$

where M is the molecular weight.

It is also possible to relate the intensity of magnetization to the rate of change of energy of the body in the magnetic field $(\partial W/\partial H)$ and we shall subsequently use the relationship

$$I = -\partial W/\partial H. \tag{1.5}$$

We can observe two fundamental types of magnetic behaviour depending on the sign of I in Equation 1.1. These are diamagnetism and paramagnetism. Phenomena such as ferro- and ferrimagnetism and antiferromagnetism are all special forms of paramagnetism.

1.3. Types of magnetic behaviour

1.3.1. DIAMAGNETISM

If the intensity of magnetization is negative, the material is said to be diamagnetic. In classical terms, the density of lines of force inside the sample is less than that outside. Such a material, when placed in an inhomogeneous magnetic field will tend to move to the region of lowest field. This repulsion from the field is perhaps best understood in terms of energies. It can be seen from Equation 1.5 that $\partial W/\partial H$ will be positive when I is negative. That is the energy of the body is *greatest* when H is greatest. The move to the region of lowest field is thus an exothermic process.

The molar susceptibility of a diamagnetic material is negative, and rather small being of the order -1 to -100×10^{-6} c.g.s. e.m.u. Diamagnetic susceptibilities do not depend on field strength, and are normally independent of temperature.

Diamagnetism is a property of all matter, and arises from the interaction of paired electrons with the magnetic field. In classical terms, electron pairs can be treated as current loops, in which case, repulsion from the field is a simple consequence of Lenz's law. In quantum terms, χ_M for a diamagnetic material is

$$\chi_M = -\frac{Ne^2}{6mc^2} \int R(r)^2 r^2 \, dr$$

where $R(r)$ is the radial part of the electronic wavefunction.

Diamagnetic susceptibilities of atoms in molecules are additive within reasonable limits, and this is of particular use in estimating the diamagnetic susceptibilities of ligand atoms and counter-ions in a transition metal complex. Much of the work of establishing the additivity of atomic susceptibilities was carried out by Pascal, and the quantities are commonly referred to as Pascal's constants. Strictly, the term applies to the values for neutral atoms – we shall use it more generally to represent any diamagnetic correction. Some typical values of these constants are listed in Table 1.1.

The diamagnetic susceptibility of a molecule may be written:

$$\chi_{mol.} = \sum_i n_i \chi_i + \sum \epsilon \tag{1.6}$$

where the molecule contains n_i atoms of atomic susceptibility (i.e. per gram atom), χ_i, and ϵ represents 'constitutive' corrections which take account of such factors as the existence of π-bonds.

There is some confusion as to sign conventions. Pascal's constants are commonly listed as $|\chi_i|$ when ϵ becomes apparently negative. In Table 1.1 we

Table 1.1. *Pascal's Constants*
(Susceptibilities per gram atom $\times 10^6$ c.g.s. e.m.u.).

Cations		Anions	
Li^+	$-1\cdot0$	F^-	$-9\cdot1$
Na^+	$-6\cdot8$	Cl^-	$-23\cdot4$
K^+	$-14\cdot9$	Br^-	$-34\cdot6$
Rb^+	$-22\cdot5$	I^-	$-50\cdot6$
Cs^+	$-35\cdot0$	NO_3^-	$-18\cdot9$
Tl^+	$-35\cdot7$	ClO_3^-	$-30\cdot2$
NH_4^+	$-13\cdot3$	ClO_4^-	$-32\cdot0$
Hg^{2+}	$-40\cdot0$	CN^-	$-13\cdot0$
Mg^{2+}	$-5\cdot0$	NCS^-	$-31\cdot0$
Zn^{2+}	$-15\cdot0$	OH^-	$-12\cdot0$
Pb^{2+}	$-32\cdot0$	SO_4^{2-}	$-40\cdot1$
Ca^{2+}	$-10\cdot4$	O^{2-}	$-12\cdot0$

Neutral Atoms			
H	$-2\cdot93$	As(III)	$-20\cdot9$
C	$-6\cdot00$	Sb(III)	$-74\cdot0$
N(ring)	$-4\cdot61$	F	$-6\cdot3$
N(open chain)	$-5\cdot57$	Cl	$-20\cdot1$
N(imide)	$-2\cdot11$	Br	$-30\cdot6$
O(ether or alcohol)	$-4\cdot61$	I	$-44\cdot6$
O(aldehyde or ketone)	$1\cdot73$	S	$-15\cdot0$
P	$-26\cdot3$	Se	$-23\cdot0$
As(V)	$-43\cdot0$		

Some common molecules			
H_2O	-13	$C_2O_4^{2-}$	-25
NH_3	-18	acetylacetone	-52
C_2H_4	-15	pyridine	-49
CH_3COO^-	-30	bipyridyl	-105
$H_2N.CH_2.CH_2NH_2$	-46	o-phenanthroline	-128

Constitutive Corrections			
C=C	$5\cdot5$	N=N	$1\cdot8$
C=C—C=C	$10\cdot6$	C=N—R	$8\cdot2$
C≡C	$0\cdot8$	C—Cl	$3\cdot1$
C in benzene ring	$0\cdot24$	C—Br	$4\cdot1$

list actual values with the correct sign. As an example, the diamagnetic contribution to the susceptibility of $K_4Fe(CN)_6$ is:

$$K^+ \ 4 \times -14 \cdot 9 = -59 \cdot 6 \times 10^{-6}$$
$$(CN)^- \ 6 \cdot x - 13 \cdot 0 = \underline{-78 \cdot 0 \times 10^{-6}}$$
$$-137 \cdot 6 \times 10^{-6} \text{ c.g.s. e.m.u.}$$

(It is conventional to ignore the diamagnetism of the iron.) As a warning against the uncritical use of Pascal's constants the measured molar susceptibility of the ligand o-phenylenebisdimethylarsine is -194×10^{-6} c.g.s e.m.u. The calculation from Pascal's constants yields

$$10 \times C = -60 \cdot 0 \times 10^{-6}$$
$$16 \times H = -46 \cdot 9 \times 10^{-6}$$
$$2 \times As(III) = -41 \cdot 8 \times 10^{-6}$$
$$\text{Correction for C in benzene ring} = \underline{\quad 1 \cdot 4 \times 10^{-6}}$$
$$-147 \cdot 3 \times 10^{-6} \text{ c.g.s.}$$

In more complicated molecules, particularly when aromatic rings are present, it is more reliable to measure the susceptibility of the ligand either free, or complexed with a diamagnetic metal.

1.3.2. PARAMAGNETISM

The intensity of magnetization of a paramagnet is positive, hence $\partial W/\partial H$ is negative and such a material will tend to move to regions of maximum field strength since this is now an exothermic process. Paramagnetic susceptibilities are positive and relatively large (χ_M in the range 100 to 100 000 $\times 10^{-6}$ c.g.s. e.m.u.). They do not depend on magnetic field strength, but usually depend very markedly on temperature.

Paramagnetism is a consequence of the interaction of orbital- and/or spin-angular momenta of unpaired electrons with the applied field. In a molecule, all atoms have a diamagnetic susceptibility, thus a measured molar susceptibility contains both paramagnetic and diamagnetic terms. If the molar susceptibility of a paramagnetic compound is corrected for the diamagnetism of all except the paramagnetic atom we obtain the susceptibility per gram atom, χ_A where

$$\chi_A = \chi_M - \chi_{mol.}$$

the term $\chi_{mol.}$ is defined in Equation 1.6.

It is conventional to ignore the underlying diamagnetism of the closed shells of the paramagnetic atom. Similarly χ_A is often labelled according to the atom being discussed, e.g. χ_{Ti} is the value of χ_A for the particular titanium complex under discussion.

The magnitude of χ_A is an inconvenient number, it is conventional to *define* the effective magnetic moment, μ, as

$$\mu = \left(\frac{3k}{N\beta^2}\right)^{\frac{1}{2}} (\chi_A T)^{\frac{1}{2}} \tag{1.7}$$

$$= 2 \cdot 828 \, (\chi_A T)^{\frac{1}{2}} \text{ Bohr magnetons (in c.g.s. units).}$$

The quantity β is the Bohr magneton:

$$\beta = \frac{eh}{4\pi mc} = 0 \cdot 927 \times 10^{-20} \text{ erg gauss}^{-1}.$$

It should be stressed that the magnetic susceptibility is the fundamental quantity and that μ is derived from it by arbitrary definition.

Principal susceptibilities

In the discussion so far, we have dealt either with a single magnetically isotropic (by assumption) body, or with single molecules. All single crystals except those belonging to the cubic class must be magnetically anisotropic and hence we must define three orthogonal *principal crystal susceptibilities.* It is normal to use the cartesian axes and define χ_x, χ_y, and χ_z as the susceptibilities with the magnetic field in those directions relative to the crystal. If the crystal has a unique rotation axis, this will be the z-axis and we can then define $\chi_{\parallel}$ and $\chi_{\perp}$ as the susceptibilities parallel and perpendicular to the z axis (not the field), i.e. $\chi_x = \chi_y = \chi_{\perp}$.

Crystal principal susceptibilities (or more easily the anisotropies, e.g. $\chi_x - \chi_y$) can be measured provided that the principal axes of the single crystal sample can be located. However, the calculations which follow in this, and subsequent chapters are referred to individual molecules. Thus it is necessary to convert from principal crystal susceptibilities to *principal molecular susceptibilities.* The conversion can be carried out only if the orientation of molecules in the cell is known, *and* if it is assumed that molecular magnetic axes coincide with the geometric axes of the molecule.

Measurements on a powdered sample will yield the average susceptibility:

$$\bar{\chi} = \frac{\chi_x + \chi_y + \chi_z}{3}$$

but note that $\bar{\mu} = \left[\dfrac{\mu_x^2 + \mu_y^2 + \mu_z^2}{3}\right]^{\frac{1}{2}}.$

1.3.3. MAGNETIC INTERACTIONS

In the above section we have assumed that the magnetic behaviour of an individual paramagnetic atom is independent of that of its neighbours, that is, the compound is *magnetically dilute*. The magnetic behaviour is then determined by a thermal distribution among available states.

If the behaviour of adjacent magnetic centres is not independent, the compound is said to be *magnetically concentrated* and its magnetic behaviour is determined by the nature of the interactions between centres. The principal types of behaviour are anti-ferromagnetism, ferromagnetism and ferrimagnetism.

(*a*) *Antiferromagnetism.* This arises if the magnetic vectors of neighbouring centres tend to couple antiparallel. In an isotropic solid this would give a diamagnetic susceptibility at 0 K. Increasing the temperature will tend to randomize the alignment and the susceptibility rises to a maximum at the Néel Point, T_N, as shown in Fig. 1.1(a). Above T_N the susceptibility falls, ultimately obeying a Curie-Weiss law, usually with positive θ (see Section 1.5.1. page 12). In anisotropic solids only χ_z falls to zero below T_N, hence the shape of the curve in Fig. 1.1(a). The susceptibility is usually only weakly field-dependent below T_N.

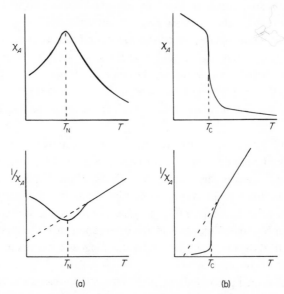

Figure 1.1. Variation of magnetic susceptibility and inverse magnetic susceptibility with temperature: (a) for an antiferromagnetic compound; (b) for a ferromagnetic compound.

Evidence for this type of alignment has been found in manganese(II) oxide by neutron diffraction. In Fig. 1.2(a) it is shown that the magnetic unit cell of MnO is twice as large as the chemical unit cell. Below T_N, an extra neutron diffraction line appears at low angle, a consequence of the interaction of the neutrons with the ordered magnetic moments of the manganese(II) ions.

Many different antiferromagnetic orderings may be postulated; two of these are shown for face-centred cubic lattices in Fig. 1.2(b) and (c).

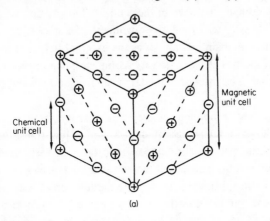

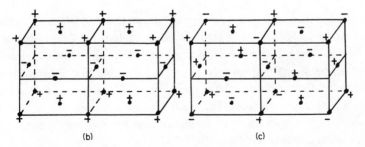

Figure 1.2. Spin ordering in antiferromagnetic lattices: (a) in manganese(II) oxide (only Mn(II) ions are shown); (b) and (c) two types of ordering in face-centred cubic lattices. The + and − refer to the two possible alignments of the magnetic vectors.

It is also possible to observe antiferromagnetic ordering over a small number of centres: this is the subject of Chapter 7.

(b) Ferromagnetism. Ferromagnetism is observed when the magnetic vectors of adjacent centres tend to allign parallel to each other. At high temperatures, the susceptibility obeys a Curie-Weiss law with negative θ. The Curie, and Curie-Weiss Laws are discussed in Section 1.5.1. As the temperature is lowered, the susceptibility increases more rapidly than expected, the prelude to an abrupt, very large, increase in susceptibility. This is illustrated in Fig. 1.1(b). The point at which $\partial\chi/\partial T$ is maximum is called the Curie-temperature, T_C. Below T_C, the

susceptibility becomes very field dependent (often $\chi \propto 1/H^2$). Many (but not all) ferromagnets have permanent moments, that is, they remain magnetized when the magnetic field is removed. The latter is a property only of ferromagnets. The unique magnetic behaviour of ferromagnets (e.g. permanent magnetization, field dependence of susceptibility), are consequences of the existence of a domain structure in the solid. Within any domain the alignment is ferromagnetic but in zero field the alignment of adjacent domains is random. The behaviour of domain boundaries as the external field is changed gives rise to the unique properties. This is a very complex phenomenon, and we cannot discuss it further. The interested reader is referred to standard physics texts on magnetism.

Surprisingly few compounds prove to be ferromagnetic: a small number of metals and alloys, a fair range of oxides, and a few transition metal halides complete the list. Yet their industrial importance is very great, ranging from permanent- and electro-magnets, to cores for inductances and magnetic recording tape and such sophisticated uses as the modulation of laser beams for speech transmission and 'bubble domain' devices for the next generation of computer stores. Studies of ferromagnets, on the other hand, produce little *chemically* significant information.

(c) *Ferrimagnetism.* Reference to Fig. 1.2 shows that the '+' and '−' spins can be considered to fall on two interpenetrating sub-lattices. Thus in the arrangement (b), in Fig. 1.2, there is ferromagnetic ordering in a layer, coupled with antiferromagnetic ordering between layers. It is possible, in a ternary compound to have two different paramagnetic atoms, and the two types can be segregated each on a particular sub-lattice. The result is that the antiferromagnetic ordering does not lead to a cancellation of moments on the two sub-lattices. Such an arrangement is said to be ferrimagnetic.

1.4. The origin of paramagnetism

Paramagnetism arises from changes in the energy levels of an atom when it is subjected to a magnetic field. We thus need to discover the new energy levels of individual atoms in the magnetic field, and then derive the susceptibility per mole of atoms from the thermal distribution of N atoms among these possible states, according to Boltzmann statistics.

The derivation of a suitable general equation for this calculation is due to J.H. Van Vleck, and it is normally referred to as Van Vleck's equation. Its derivation is given in reference 4, page 181, for an electric susceptibility. The derivation for magnetic susceptibility proceeds similarly. We give the derivation here particularly in order to comment upon the approximations made in it.

We first assume that there is no field dependence of the susceptibility, that is, we are dealing *only* with magnetically dilute systems. Secondly, it is assumed that the energy of the ith level of the atom may be expressed as a power series in the magnetic field:

$$W_i = W_i^0 + W_i^{(1)}H + W_i^{(2)}H^2 + \ldots \ldots \tag{1.8}$$

It will become apparent as we proceed that the series must terminate at the second-order term, otherwise the final expression will contain H, a possibility which we excluded above.

In Equation 1.8, W_i^0 is the energy of the level in the absence of the field. $W_i^{(1)}$ and $W_i^{(2)}$ are the first- and second-order Zeeman coefficients respectively (it should be noted that these do not have dimensions of energy). A finite first-order Zeeman coefficient would give rise, e.g., to the well-known Zeeman splitting in a magnetic field of the sodium atomic emission lines. For a given atom

$$\begin{aligned} I_i &= -\partial W_i/\partial H \\ &= -W_i^{(1)} - 2W_i^{(2)}H \end{aligned} \tag{1.9}$$

and its susceptibility, χ_i, is

$$\chi_i = I_i/H.$$

Thus, for N atoms (i.e. one mole of atoms), we need to sum Equation 1.9 over the N atoms, distributed over the states W_i according to Boltzmann statistics:

$$\chi_A = \frac{\dfrac{N}{H} \sum_i [-W_i^{(1)} - 2W_i^{(2)}H] \exp(-W_i/kT)}{\sum_i \exp(-W_i/kT)} \tag{1.10}$$

We now assume that any splitting of energy levels due to a first order Zeeman effect is small compared with kT, the thermal energy available to the system. This assumption is justifiable at room temperature since the first order Zeeman splittings are a few wave-numbers for normal laboratory fields, whereas kT is about two hundred wave-numbers. At very low temperatures, the assumption will not be valid and calculations must be made from Equation 1.10.

However, if $W_i^{(1)} \ll kT$, the exponentials in (1.10) may be simplified, since $\exp(-x) \simeq (1-x)$ if x is small.

Thus we may write, using (1.8)

$$\exp(-W_i/kT) = \exp(-W_i^0/kT)[(1 - H.W_i^{(1)}/kT)(1 - H^2 W_i^{(2)}/kT)(\ldots)].$$

Thus Equation 1.10 becomes

$$\chi_A = \frac{\dfrac{N}{H} \sum_i [-W_i^{(1)} - 2W_i^{(2)}H][\exp(-W_i^0/kT)][1 - W_i^{(1)}H/kT][1 - W_i^{(2)}H^2/kT]}{\sum_i [\exp(-W_i^0/kT)][1 - W_i^{(1)}H/kT][1 - W_i^{(2)}H^2/kT]}$$

Multiplying out, and ignoring any terms containing H gives

$$\chi_A = \frac{N \sum_i (W_i^{(1)^2}/kT - 2W_i^{(2)}) \exp(-W_i^0/kT)}{\sum_i \exp(-W_i^0/kT)} \tag{1.11}$$

Equation 1.11 is Van Vleck's susceptibility equation.

Thus the calculation of χ_A depends upon the evaluation of W_i^0, $W_i^{(1)}$, and $W_i^{(2)}$. This is dealt with in detail in Chapters 3–5.

1.5. Generalizations about the magnetic behaviour of d-transition metal complexes

1.5.1. SIMPLIFICATION OF VAN VLECK'S EQUATION

It is possible to simplify Van Vleck's equation (1.11) in various circumstances. We do this here because it allows us to understand the origin of the various generalizations about magnetic behaviour in transition metal complexes, especially the temperature dependence of magnetic moments, and its relationship to structure.

In order to do this, we must anticipate certain of the results of Chapters 2 and 3. The first order Zeeman coefficients are integrals of type $\int \psi_i^* \hat{\mu} \psi_i \, d\tau$ where ψ_i are the wave-functions of a level which is degenerate in zero field; $\hat{\mu}$ is the appropriate magnetic moment operator. If $W_i^{(1)}$ is finite, the level i is split into a set of equally spaced components separated by $g\beta H$ (see Chapter 4). The susceptibility arises from the changes in energy on thermal population of these new states. If the second-order Zeeman coefficients are finite there will be mixing of the wave-functions of the ith and jth levels (i.e. levels not degenerate in zero field). It is a sum of terms of the type

$$\frac{\int \psi_i^* \hat{\mu} \psi_j \, d\tau . \int \psi_j^* \hat{\mu} \psi_i \, d\tau}{W_i^0 - W_j^0}.$$

The purturbation theory requires that this term is not included when $W_i^0 = W_j^0$. A finite second order Zeeman effect causes an equal lowering of the energy of the levels i and a corresponding raising of those of levels j, but does not produce any changes in degeneracy.

It is worth noting that $W_i^0 - W_j^0$ is negative hence $W_i^{(2)}$ is negative and its contribution to χ_A in Equation 1.11 is positive (i.e. paramagnetic). If there are more than two non-degenerate levels in zero field, then all possible interactions must be considered. See, for example, reference 5, page 269.

We are now in a position to simplify Van Vleck's equation. There are four general situations.

(A) *In zero field, there is only one energy level, and this is degenerate.*

In this case $W_i^{(2)} = 0$ since there is no set of levels j with which to interact. Further, our reference energy level is W_i^0, which we may put as zero, thus $\exp(-W_i^0/kT) = 1$. Equation 1.11 then becomes:

$$\chi_A = \frac{N \sum_i W_i^{(1)^2}/kT}{n} \tag{1.12}$$

where n is the degeneracy of the level. For a particular case, $\sum_i W_i^{(1)^2}$ is a constant, and we can write (1.12) as

$$\chi_A = C/T \tag{1.13}$$

where C is a constant.

Equation 1.13 is the Curie law. Since we have defined μ in Equation 1.7 as

$$\mu = 2 \cdot 828 (\chi_A T)^{\frac{1}{2}}$$

it follows that if a Curie law is obeyed

$$\mu = 2 \cdot 828 (C)^{\frac{1}{2}}.$$

That is, μ is independent of temperature if the susceptibility obeys a Curie law. Further, in this case, a plot of $1/\chi_A$ *vs.* T is straight line through the origin.

Although the Curie law may be approximately obeyed in many cases, no transition metal complex exists which has only one energy level in zero-field. Thus case (A) has no direct application.

(B) *The ground term is a singlet, and there is at least one degenerate excited state but these are all $\gg kT$ above the ground state.*

In this case $W_i^{(1)}$ is zero since the level i is a singlet, once again we can put $W_i^0 = 0$ *i.e.* $\exp(-W_i^0/kT) = 1$. Further, although $W_j^{(1)}$ may well be finite, $\exp(-W_j^0/kT)$ will be approximately zero since $W_j^0 \gg kT$ (i.e. the level j is not thermally populated) and so there will be no first-order Zeeman contribution to χ_A. The only finite term in Equation 1.12 is thus $W_i^{(2)}$ and in this case

$$\chi_A = N \sum_i W_i^{(2)} = N\alpha$$

The form of $W_i^{(2)}$ is given above: it only depends on the separation $(W_i^0 - W_j^0)$ hence in this situation χ_A is a constant (independent of temperature). This is the origin of 'temperature independent paramagnetism' – (T.I.P.). It is common to represent the T.I.P. term by the constants $N\alpha$. It must be stressed that when a T.I.P. is observed, it is χ_A which does not depend on temperature. The magnetic moment derived from a T.I.P. will be a function of $\sqrt{T}$.

This a real situation, and is the origin of the paramagnetism of such d^0 systems as the permanganate or chromate ions. In most cases $(W_i^0 - W_j^0)$ is rather large, hence the T.I.P. is small, typically $\sim100 \times 10^{-6}$ c.g.s. In other cases, the T.I.P. may be quite large (see e.g. octahedral d^1 systems in Chapter 4). It may be shown that as $(W_i^0 - W_j^0)$ becomes smaller the T.I.P. behaviour merges smoothly into a Curie law behaviour: see e.g. reference 5, page 255.

(C) *The ground state is degenerate and all excited states are $\geqslant kT$ above this.*

This is simply the sum of cases (A) and (B), thus

$$\chi_A = C/T + N\alpha. \tag{1.15}$$

We have seen that $N\alpha$ will be small, further when T is small, the Curie law contribution will be very large. Hence, at low temperatures a Curie law will be obeyed: a plot of $1/\chi_A$ *vs.* T will be a straight line through the origin, but slightly concave towards the T axis at high temperatures. The magnetic moment will be independent of temperature, apart from small increases at higher temperatures.

This is a real situation and will be observed in any complex of symmetry O_h or T_d for which the ground term is of A or E symmetry. This is because the first excited term will be $10Dq$ above the ground term (say 10 000 cm^{-1}), and as we shall see in Chapter 2, the ground terms are not split by spin–orbit coupling so

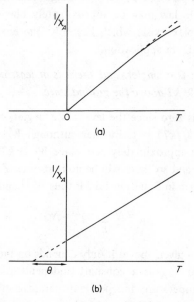

(a)

(b)

Figure 1.3. Inverse magnetic susceptibility *vs.* temperature for transition complexes: (a) for an ion with an A or E ground term; (b) for an ion with a T ground term.

that there is no low-lying excited state. This behaviour is illustrated in Fig. 1.3(a). An approximate value for $N\alpha$ can be obtained by finding the constant which must be subtracted in order to bring the $1/\chi_A$ vs. T plot to a straight line.

(D) *The ground term is degenerate and there are thermally accessible degenerate excited terms.*

In this case, we can still put $W_i^0 = 0$, i.e. $\exp(-W_i^0/kT) = 1$, but now there is the possibility that both $W_j^{(1)}$ and $W_j^{(2)}$ are finite as well as $W_i^{(1)}$ and $W_i^{(2)}$ and in this case $\exp(-W_j^0/kT)$ is not negligible. (There may well be further levels, k etc., giving rise to $W_k^{(1)}$, etc.) In this case, no simplification of Van Vleck's equation is possible, and we may expect a complicated temperature dependence of the susceptibility. In practice, it is often found that the susceptibility obeys a Curie-Weiss law:

$$\chi_A = C/(T + \theta) \tag{1.15}$$

where θ is a constant (the Weiss constant).

In these cases, the magnetic moment will depend on temperature. If (1.15) is obeyed, a plot of $1/\chi_A$ vs. T will be a straight line, but with a finite intercept on the temperature axis. This situation is illustrated in Fig. 1.3(b). It should be stressed that conformity to the Curie-Weiss law in these cases is coincidental and no particular significance attaches to a compound which does not obey this law.

There is considerable misunderstanding of the significance of the Weiss constant, θ. This was originally introduced by Weiss as part of his molecular field treatment of anti-ferromagnetism. It is true that, in most cases, at sufficiently high temperatures the magnetic susceptibilities of ferromagnets and antiferromagnets obey Curie-Weiss laws with θ, respectively negative and positive (writing the law as in Equation 1.15). These signs will be reversed if the expression is written

$$\chi_A = C/(T - \theta)$$

as used by some workers and this is the source of some confusion. Further, in a magnetically concentrated compound there is some relationship between the magnitude of θ and the strength of the magnetic interaction. However, it does not follow that if θ is finite some magnetic interaction exists. Thus it is shown in Chapter 4 that isolated regular octahedral d^1 complexes obey a Curie-Weiss law with $\theta \simeq 200$ K. Some authors calculate μ from the expression

$$\mu = 2 \cdot 828 \, [\chi_A(T + \theta)]^{\frac{1}{2}}.$$

This procedure is never justified for a magnetically dilute compound where its use only serves to conceal the truth about the magnetic behaviour of the compound. The quantity μ is *defined* by Equation 1.7 and this should always be used. Although the inclusion of θ may be justified in interacting systems, it is very doubtful whether it confers any advantage.

The behaviour predicted in case (D) may be expected for any transition metal complex which has a T symmetry ground term. As is shown in Chapter 2, these terms are split by spin–orbit coupling interactions. The resulting states (usually degenerate) are separated by a few hundred wave-numbers and thus will be thermally populated. These compounds will have temperature dependent magnetic moments. Table 1.3 lists the ground terms for all d^n configurations in O_h and T_d symmetry crystal fields. With the exception of d^5, for a given configuration the ground state is of symmetry T in O_h when it is A or E in T_d and vice-versa. Thus cases (C) and (D) are the basis of the much-used criterion of stereochemistry based upon the temperature dependence or otherwise of the magnetic moment. We shall see in Chapter 5 that this criterion must be used with extreme caution.

1.5.2. THE MAGNITUDE OF MAGNETIC MOMENTS

It is possible to make some broad generalizations about the magnitude of magnetic moments, based upon the nature of the crystal field ground term. The common starting point for this discussion is to state the expression for a free ion with ground state quantum numbers L and S, i.e.

$$\mu = [L(L + 1) + 4S(S + 1)]^{\frac{1}{2}} \text{ Bohr magnetons.} \qquad (1.16)$$

It is instructive to derive this expression because in our experience the reader will readily accept that the magnetic susceptibility derived *via* Van Vleck's equation is the result of changes in molecular energy levels of the ion in the magnetic field. However, the connection between this and the quantum numbers which define the energies of the terms is not so readily appreciated.

We have already noted that the first order Zeeman energies are calculated from integrals of the type $\int \psi_i^* \hat{\mu} \psi_i d\tau$ where the ψ_i's are all of the wave-functions of the ground term. Assuming that $W_i^{(2)}$ is zero, and that the system is magnetically isotropic we can use the magnetic moment operator

$$\hat{\mu}_z = (\hat{L}_z + 2\hat{S}_z)\beta H$$

It is shown in Chapter 3, that if the wave-functions ψ_i are defined by their M_L and M_S, then when $M_L = M_L'$ and $M_S = M_S'$

$$\int \psi_i^*(M_L, M_S)\hat{\mu}_z \psi_i(M_L', M_S') d\tau = (M_L + 2M_S)\beta H \qquad (1.17)$$

but is zero when $M_L \neq M_L'$ or $M_S \neq M_S'$.

Thus for a free ion with quantum numbers L and S we shall generate a set of states in the magnetic field, each with energy specified by (1.17). These energies depend upon M_L and M_S but we can write the allowed values of M_L as $L, L - 1$,

... $-L$ and similarly for M_S. We shall then have integrals in (1.17) of the form

$[L + 2S]\beta H$, i.e. $W_i^{(1)} = [L + 2S]\beta$,

$[(L - 1) + 2S]\beta H$, i.e. $W_i^{(1)} = [(L - 1) + 2S]\beta$,

$[(L - 1) + 2(S - 1)]\beta H$, etc., using all combinations from L to $-L$ and S to $-S$.

Assuming that $W_i^{(2)}$ is zero, and setting $W_i^0 = 0$, we can write (1.11) as

$$\chi_A = \frac{N \sum_i (W_i^{(1)^2}/kT)}{\sum_i \exp(-W_i^0/kT)} \qquad (1.18)$$

In (1.18)

$$\sum_i \exp(-W_i^0/kT) = (2L + 1)(2S + 1) \qquad (1.19)$$

since each exponential is unity.

We can write down the terms in the numerator of (1.18) systematically as follows:

$$\sum_i \frac{W_i^{(1)^2}}{kT} =$$

$$\frac{\beta^2}{kT} \left\{ \begin{array}{l} [L + 2S]^2 \quad + [(L - 1) + 2S]^2 + [(L - 2) + 2S]^2 \ldots + [(-L) + 2S]^2 \\ +[L + 2(S - 1)]^2 + \ldots \\ +[L + 2(S - 2)]^2 + \ldots \\ \quad \cdot \\ \quad \cdot \\ \quad \cdot \\ +[L + 2(-S)]^2 \quad + \ldots \qquad\qquad\qquad\qquad + [(-L) + 2(-S)]^2 \end{array} \right\} \qquad (1.20)$$

If we evaluate these terms in suitable pairs we can eliminate cross terms, $(L \times S)$, as follows: taking the first and last members of the first column in (1.20) we have

$$[L + 2S]^2 + [L - 2S]^2 = 2[L^2 + 4S^2],$$

similarly the second and penultimate terms are

$$[L + 2(S - 1)]^2 + [L - 2(S - 1)]^2 = 2[L^2 + 4(S - 1)^2]$$

and so on. The terms in the curly bracket of Equation 1.20 thus reduce to series of sums of squares:

$$2[L^2 + (L - 1)^2 + (L - 2)^2 \ldots 0^2]$$

with *each* term in this series occurring $(2S + 1)$ times and $2[S^2 + (S - 1)^2 + (S - 2)^2 \ldots 0^2]$ with each term occurring $(2L + 1)$ times.

Equation 1.20 thus reduces to:

$$\sum_i W_i^{(1)^2}/kT = \beta^2/kT\{2(2S + 1)[L^2 + (L - 1)^2 + \ldots 0^2]$$
$$+ 2(2L + 1)[4S^2 + 4(S - 1)^2 + \ldots 4 \times 0^2]\} \qquad (1.21)$$

but the sum of the squares of the first n natural numbers is

$$S_n = \frac{n(n + 1)(2n + 1)}{6}.$$

Thus (1.21) becomes

$$\sum_i W_i^{(1)^2}/kT = \frac{2\beta^2}{6kT}\{(2S + 1)(L)(L + 1)(2L + 1)$$
$$+ 4(2L + 1)(S)(S + 1)(2S + 1)\}. \qquad (1.22)$$

Combining (1.22) and (1.19) into (1.18) gives

$$\chi_A = \frac{2N\beta^2}{6kT}\left\{\frac{L(L + 1)(2S + 1)(2L + 1) + 4S(S + 1)(2L + 1)(2S + 1)}{(2S + 1)(2L + 1)}\right\}$$

$$= \frac{N\beta^2}{3kT}[L(L + 1) + 4S(S + 1)]$$

but

$$\mu = [3k/N\beta^2]^{\frac{1}{2}}[\chi_A T]^{\frac{1}{2}}$$
$$= [L(L + 1) + 4S(S + 1)]^{\frac{1}{2}}$$

It is perhaps rather difficult to follow this derivation so let us put in the relevant numbers for a ^4P term $(L = 1, S = \frac{3}{2})$. M_L will be 1, 0 or -1 and M_S will be $\frac{3}{2}, \frac{1}{2}, -\frac{1}{2}$, or $-\frac{3}{2}$. Equation 1.20 is thus

$$\sum_i W_i^{(1)^2}/kT = \beta^2/kT \left\{ \begin{array}{l} (a)(1 + 2 \times \frac{3}{2})^2 + (b)(0 + 2 \times \frac{3}{2})^2 + (c)(-1 + 2 \times \frac{3}{2})^2 \\ + (d)(1 + 2 \times \frac{1}{2})^2 + (e)(0 + 2 \times \frac{1}{2})^2 + (f)(-1 + 2 \times \frac{1}{2})^2 \\ + (g)(1 - 2 \times \frac{1}{2})^2 + (h)(0 - 2 \times \frac{1}{2})^2 + (i)(-1 - 2 \times \frac{1}{2})^2 \\ + (j)(1 - 2 \times \frac{3}{2})^2 + (k)(0 - 2 \times \frac{3}{2})^2 + (l)(-1 - 2 \times \frac{3}{2})^2 \end{array} \right\} \quad (1.23)$$

The resultant splitting in the magnetic field is shown in Fig. 1.4.
The letters in parentheses serve to identify the terms.
Taking terms in pairs:

(a) + (j) = $2(1 + 4 \times \frac{9}{4})$; (b) + (k) = $2(0 + 4 \times \frac{9}{4})$; (c) + (l) = $2(1 + 4 \times \frac{9}{4})$;
(d) + (g) = $2(1 + 4 \times \frac{1}{4})$; (e) + (h) = $2(0 + 4 \times \frac{1}{4})$; (f) + (i) = $2(1 + 4 \times \frac{1}{4})$.

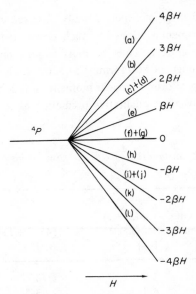

Figure 1.4. Splitting of a ^4P term in a magnetic field. The letters in parentheses refer to the terms in Equation 1.23 giving rise to each level.

Adding all of these together gives a total of 68. Hence, from Equation 1.18

$$\chi_A = \frac{N\beta^2}{kT} \times \frac{68}{3 \times 4} = \frac{N\beta^2}{3kT} \times 17$$

and

$$\mu = (17)^{\frac{1}{2}}, \text{ i.e. } [L(L+1) + 4S(S+1)]^{\frac{1}{2}}.$$

If spin–orbit coupling is included, similar expressions may be derived for a state specified by J, namely

$$\mu = g[J(J+1)]^{\frac{1}{2}}$$

where

$$g = \frac{3}{2} + \frac{S(S+1) - L(L+1)}{2J(J+1)}.$$

The derivation of these expressions is given in, for example, reference 2, page 25.

Quenching of orbital angular momentum

The values of μ calculated from Equation 1.16 are listed in Table 1.2 for all d-configurations. We also give there values of μ calculated from the so-called 'spin-only' Equation 1.24

$$\mu_{s.o.} = [4S(S+1)]^{\frac{1}{2}} \tag{1.24}$$

as well as the general range of experimentally observed values. It is clear from this Table that experimental values lie much closer to those calculated from (1.24) than to Equation 1.16. Thus we must enquire how the orbital contribution in a free ion is quenched on forming a complex. The quantum-mechanical basis of the quenching of orbital angular momentum is demonstrated in Chapter 4, but there is a convenient qualitative prediction as to when orbital contributions will appear.

Table 1.2. *Magnetic moments of first row transition metal spin-free configurations.*

No. of d electrons	L	S	Free ion ground term	$\mu = [4S(S+1) + L(L+1)]^{\frac{1}{2}}$ B.M.	$\mu_{s.o.} = [4S(S+1)]^{\frac{1}{2}}$ B.M.	μ observed at 300 K
1	2	$\frac{1}{2}$	^2D	3·00	1·73	1·7–1·8
2	3	1	^3F	4·47	2·83	2·8–2·9
3	3	$\frac{3}{2}$	^4F	5·20	3·87	3·7–3·9
4	2	2	^5D	5·48	4·90	4·8–5·0
5	0	$\frac{5}{2}$	^6S	5·92	5·92	5·8–6·0
6	2	2	^5D	5·48	4·90	5·1–5·7
7	3	$\frac{3}{2}$	^4F	5·20	3·87	4·3–5·2
8	3	1	^3F	4·47	2·83	2·9–3·9
9	2	$\frac{1}{2}$	^2D	3·00	1·73	1·7–2·2
10	0	0	^1S	0·00	0·00	0

The argument is as follows: For an electron in a particular orbital to have orbital angular momentum about a specified axis, it must be possible, by rotation about that axis, to convert the orbital into an *identical, degenerate* orbital which has a vacancy for an electron of the spin concerned. Thus in a free ion the orbital d_{xy} can be converted into $d_{x^2-y^2}$ by a 45° rotation and d_{xz} into d_{yz} by a 90° rotation both about the z-axis. Orbital contributions may arise from electrons in any of these orbitals. But in an octahedral or tetrahedral crystal field, the orbitals of $t_{2(g)}$ symmetry (d_{xy}, d_{yz} and d_{xz}) are no longer degenerate with the $e_{(g)}$ orbitals ($d_{x^2-y^2}$ and d_{z^2}) hence the orbital contribution about z arising from the d_{xy}, $d_{x^2-y^2}$ pair of orbitals now disappears. Further, since $d_{x^2-y^2}$ and d_{z^2} cannot be interconverted by rotation there is no orbital contribution associated with this pair of orbitals, hence their description as a 'non-magnetic doublet', i.e. no *orbital* magnetism.

However, the $t_{2(g)}$ orbitals may be interconverted by rotation about suitable axes and so we may expect orbital contributions from $t_{2(g)}$ electrons. Thus in O_h symmetry, d^1 and d^2 will have orbital contributions but d^3 will not because any attempt at rotation will put two electrons of the same spin in a single orbital, contrary to the exclusion principle. Table 1.3 summarizes the results of these considerations.

Table 1.3. *Quenching of the orbital contribution, to the magnetic moment, due to the ligand field.*

		Stereochemistry					
		Octahedral			Tetrahedral		
No. of d-electrons	Free ion ground term	$t_{2g}^n e_g^m$ ground configuration	ligand field ground term	Quenching of orbital contribution*	$e^n t_2^m$ ground configuration	ligand field ground term	Quenching of orbital contribution*
1	^2D	t_{2g}^1	$^2T_{2g}$	No	e^1	^2E	Yes
2	^3F	t_{2g}^2	$^3T_{1g}$	No	e^2	$^3A_2$	Yes
3	^4F	t_{2g}^3	$^4A_{2g}$	Yes	$e^2 t_2^1$	$^4T_1$	No
4	^5D	$t_{2g}^3 e_g^1$	^5E_g	Yes	$e^2 t_2^2$	$^5T_2$	No
		t_{2g}^4	$^3T_{1g}$	No	–	–	–
5	^6S	$t_{2g}^3 e_g^2$	$^6A_{1g}$	Yes	$e^2 t_2^3$	$^6A_1$	Yes
		t_{2g}^5	$^2T_{2g}$	No	–	–	–
6	^5D	$t_{2g}^4 e_g^2$	$^5T_{2g}$	No	$e^3 t_2^3$	^5E	Yes
		t_{2g}^6	$^1A_{1g}$	Yes	–	–	–
7	^4F	$t_{2g}^5 e_g^2$	$^4T_{1g}$	No	$e^4 t_2^3$	$^4A_2$	Yes
		$t_{2g}^6 e_g^1$	^2E_g	Yes	–	–	–
8	^3F	$t_{2g}^6 e_g^2$	$^3A_{2g}$	Yes	$e^4 t_2^4$	$^3T_1$	No
9	^2D	$t_{2g}^6 e_g^3$	^2E_g	Yes	$e^4 t_2^5$	$^2T_2$	No

* Is *complete* quenching of the orbital contribution expected?

It will be noted in Table 1.3 that those configurations giving rise to orbital contributions to μ also have orbital triplet (T_1 or T_2) ground terms: this is not coincidence and is discussed in Chapter 4.

The above is the origin of the 'spin-only' magnetic moment. Since each electron has $s = \frac{1}{2}$ it is easy to see that for n *unpaired* electrons, Equation 1.24 becomes

$$\mu_{s.o.} = [n(n + 2)]^{\frac{1}{2}} \text{ Bohr magnetons.}$$

Thus it is possible in principle to derive the number of unpaired electrons in a complex from its magnetic moment.

It is usually more convenient to discuss magnetic properties in terms of μ, both because it is a conveniently small number, and also because it is easier to see whether μ depends on temperature rather than the detail of how χ_A depends on temperature.

Concluding remarks

In the preceding sections of this chapter, certain general qualitative statements have been made about the magnitude and temperature dependence of magnetic moments of octahedral and tetrahedral transition metal complexes. We shall continue to restrict the discussion to complexes with these 'cubic' or axially distorted cubic symmetries. However, the same approach can be used in other symmetries. We shall divide the compounds into two general classes below: unfortunately this is the limit to many chemists' approach to magnetism. The remaining chapters of this book exist because this is not the limit to the understanding of the subject and also because a deeper understanding reveals the weaknesses of the generalizations.

Transition metal complexes of cubic symmetry may be divided into two general classes as regards their magnetic properties:

Class (1) those for which the ground crystal field term has A or E symmetry;
Class (2) those in which the ground term has T symmetry.

Ground terms are listed in Table 1.3.

Class 1. A or E crystal field ground terms: The ground term is not split by spin-orbit coupling and so the first excited term will be at much higher energy and will be a crystal field term. We thus have the situation (C) in Section 1.5.1, i.e.

$$\chi_A = \frac{C}{T} + N\alpha$$

Thus a plot of $1/\chi_A$ *vs.* T will be a straight line through the origin except for some curvature at high temperatures. The magnetic moment will be independent of temperature at low temperatures and have the spin-only value. It is shown in Chapter 4 that second order spin–orbit coupling mixes the wave-functions of the orbital triplet excited state into those of the ground state. This changes the value of μ to

$$\mu = \mu_{s.o.} \left(1 - \frac{\alpha\lambda}{10Dq} \right), \tag{1.25}$$

where λ and Dq are spin–orbit coupling and crystal field parameters. α is 4 for A terms and 2 for E terms. The parameter λ is positive for less than half-filled shells and negative for more than half-filled shells. Hence μ will be less than spin-only for d^1–d^4 and greater than spin-only for more than 5 electrons.

It can also be shown that the T.I.P. term is

$$N\alpha = \frac{\gamma N\beta^2}{10Dq}$$

where γ is 8 or 4 for A or E terms respectively.

Class 2. Compounds with orbital triplet ground terms may have magnetic moments very different from the spin-only value, and they will generally be temperature dependent. Departures from the spin-only moment may be very great, for example a few nickel(II) complexes have been reported as having $\mu > 4$ B.M. As a less extreme example we may consider the properties of the tetrahedral nickel(II) complex $(Et_4N)_2NiCl_4$ for which $\mu = 3 \cdot 8$ B.M. at room temperature. There will be two unpaired electrons thus $\mu_{s.o.} = 2 \cdot 83$ B.M. The observed value is close to that of $\mu_{s.o.}$ for three unpaired electrons, thus it is not clear whether we are considering a d^8 configuration with large orbital contribution or d^7 with none. However, in T_d the ground terms are $^3T_1$ or $^4A_2$ respectively, and the measured μ for the compound decreases markedly with temperature thus the compound has the $^3T_1$ ground term. The analogous iodide has $\mu = 3 \cdot 2$ B.M. at room temperature with very little temperature dependence. Thus the case is less clear here since octahedral nickel(II) complexes commonly have $\mu \simeq 3 \cdot 1$ B.M. This is because $|\lambda|$ is fairly large and Dq relatively small for tetrahedral nickel(II) so that the predicted value of μ in Equation 1.25 is significantly greater than the spin-only value. This example serves to underline the care which must be taken when using magnetic properties as a criterion of structure.

REFERENCES

1. L.F. Bates (1961). *Modern Magnetism*, Cambridge University Press.
2. A. Earnshaw (1968). *Introduction to Magnetochemistry*, Academic Press, London.
3. B.N. Figgis and J. Lewis (1960). In *Modern Co-ordination Chemistry*, Eds. J. Lewis and R.J. Wilkins, Interscience, New York.
4. J.H. Van Vleck (1965). *The Theory of Electric and Magnetic Susceptibilities*, Oxford University Press.
5. B.N. Figgis (1966). *Introduction to Ligand Fields*, Interscience, New York.

<table>
<tr><td>2</td><td>Free ions and crystal field theory</td></tr>
</table>

2.1. Introduction

The calculation of magnetic properties effectively takes over where crystal field theory leaves the calculation of energy levels in a transition metal complex. We thus need the results of crystal field theory in terms of the energies and wave-functions of crystal field terms, but it is not within the province of this book to develop the theory. Fortunately, many texts exist which deal very adequately with crystal field theory and the related ligand field theory at all levels from the fairly elementary, e.g. Sutton [1], to the very complex, e.g. that of Griffiths [7]. A selection of books are listed in the bibliography, see e.g. references [1–7]. This chapter sets out the results that we shall require subsequently, but provides only the minimum of discussion. There are many very fascinating problems in this area, but space forces us to ignore them.

The terms crystal field and ligand field are used essentially interchangeably by most people. There is a difference in approach between these two, and strictly speaking they should not be interchanged. The crystal field theory was developed by Bethe [8] in an attempt to perform calculations on ionic crystals. It is thus concerned with an ionic model, and seeks to develop the electrostatic potential at a particular ion by considering all other ions as point charges. The crystal field theory does not enquire how the point charges became arranged in the particular pattern, nor about the nature of the bonds holding the ions in position. Clearly a transition metal complex is not held together by purely ionic bonds: this is one of the weaknesses implied in the use of the crystal field theory. The ligand field theory uses a molecular-orbital approach to the bonding and electronic structure in the complex, and is thus potentially better able to describe a transition metal complex. It would appear that the latter theory is to be favoured, but unfortunately it is difficult to carry out the necessary calculations. Even in the crystal field theory, very few *ab initio* calculations of the crystal field parameter Dq have been carried out, and these with only little success in a d^1 system. The difficulty here is uncertainty about precise radial wave-functions, but a properly parameterized crystal field calculation gives precise answers in terms of parameters which are usually experimentally accessible. It is possible in a crystal field calculation to make allowance for the effects of covalent bonding, and we shall do so by the introduction of further parameters. The main difficulty then lies in the interpretation of the experimental values of these latter parameters.

We shall first consider the energies of free ions and follow this by a discussion of the crystal field theory of 'cubic' complexes (i.e. those in which the crystal field has symmetry O_h or T_d). These results will be required in Chapter 4. Finally, we shall make some comment upon the energy levels in axially distorted complexes, i.e. those with symmetry derived by distorting a 'cubic' field complex along a four-fold or three-fold axis.

2.2. Free ions

2.2.1. INTERELECTRON REPULSION AND FREE ION TERMS

The energy levels of a free ion are determined by the possible combinations of the quantum numbers m_l and m_s of the individual electrons in the configuration. Any closed shell of electrons may be ignored in most instances because its effect is to uniformly raise the energies of all terms in the manifold. We are usually interested in the relative energies of two terms rather than their absolute values.

For a particular configuration we can find possible values for the quantum numbers L and S since L is the maximum value of Σm_1 for a given value of S (= maximum value of Σm_s). The allowed values of M_L and M_S are then:

$$M_L = L, (L-1), \ldots -L,$$
$$M_S = S, (S-1), \ldots -S.$$

We shall restrict the discussion to incomplete d-shells. It is readily seen that for a d^1 ion there are ten possible combinations of m_l and m_s and all of these will have the same energy since there is only one electron. The maximum Σm_l is 2, so that the value of L for this ten-fold degenerate term is also two. The only free ion term for d^1 is thus ^2D (the term is labelled ^{2S+1}X where X is S, P, D, F, $G \ldots$ when $L = 0, 1, 2, 3, 4 \ldots$).

Interelectron repulsion

When there is more than one d-electron, their mutual repulsions will contribute to the possible energies. Two electrons with the same m_1 (i.e. in the same orbital) might be expected to repel one another more strongly than if they were in different orbitals.

Table 2.1. *Free ion terms arising from the d^2 configuration.*

Term	Energy		Separation from ground term
	Condon–Shortley	Racah	
^3F	$F_0 - 8F_2 - 9F_4$	$A - 8B$	0
^1D	$F_0 - 3F_2 + 36F_4$	$A - 3B + 2C$	$5B + 2C$
^3P	$F_0 + 7F_2 - 84F_4$	$A + 7B$	$15B$
^1G	$F_0 + 4F_2 + F_4$	$A + 4B + 2C$	$12B + 2C$
^1S	$F_0 + 14F_2 + 126F_4$	$A + 14B + 7C$	$22B + 7C$

If we consider the d^2-configuration, there are 45 ways of arranging the two electrons in the d-orbitals which do not contravene the exclusion principle. [The number of arrangements is $n!/m!(n - m)!$ for m electrons; n is ten for d-orbitals to allow for $m_s = \pm\frac{1}{2}$]. There are several ways of finding the L and S of the resulting free ion terms – they are illustrated for example by Ballhausen [2] and Figgis [3]. The resulting terms for d^2 are listed in Table 2.1. The total

Table 2.2. *Weak field terms arising from free ion ground terms and from other free ion terms with the same spin multiplicity.*

Configuration	Free ion terms	Weak field terms*	
		in O_h	in T_d
d^0	^1S	$^1A_{1g}$	$^1A_1$
d^1	^2D	$^2T_{2g}, \, ^2E_g$	$^2E, \, ^2T_2$
d^2	^3F	$^3T_{1g}, \, ^3T_{2g}, \, ^3A_{2g}$	$^3A_2, \, ^3T_2, \, ^3T_1$
	^3P	$^3T_{1g}$	$^3T_1$
d^3	^4F	$^4A_{2g}, \, ^4T_{2g}, \, ^4T_{1g}$	$^4T_1, \, ^4T_2, \, ^4A_2$
	^4P	$^4T_{1g}$	$^4T_1$
d^4	^5D	$^5E_g, \, ^5T_{2g}$	$^5T_2, \, ^5E$
d^5	^6S	$^6A_{1g}$	$^6A_1$
d^6	^5D	$^5T_{2g}, \, ^5E_g$	$^5E, \, ^5T_2$
d^7	^4F	$^4T_{1g}, \, ^4T_{2g}, \, ^4A_{2g}$	$^4A_2, \, ^4T_2, \, ^4T_1$
	^4P	$^4T_{1g}$	$^4T_1$
d^8	^3F	$^3A_{2g}, \, ^3T_{2g}, \, ^3T_{1g}$	$^3T_1, \, ^3T_2, \, ^3A_2$
	^3P	$^3T_{1g}$	$^3T_1$
d^9	^2D	$^2E_g, \, ^2T_{2g}$	$^2T_2, \, ^2E$
d^{10}	^1S	$^1A_{1g}$	$^1A_1$

* In order of increasing energy.

degeneracy of all terms is 45, as it must be. Hund's rules enable us to find the ground term since they require that this has the largest S. If more than one term satisfies this criterion, that with largest L lies lowest. Thus in d^2, ^3F is the ground term. Table 2.2 gives the ground term for all d-configurations, and any excited term with the same multiplicity as the ground term.

Hund's rules only tell us the ground term and not the ordering of the excited terms, e.g. in d^2, ^1D lies below the other spin–triplet (^3P). Assuming C/B is 4·8 (see Table 2.3), the ^1D is at 14·6B compared with 15B for the ^3P. It is, of course, necessary to calculate the energies of the free ion terms. The absolute calculation fails because we do not know the radial wave-function of the d-orbitals with sufficient precision, and the solution is to introduce interelectron repulsion parameters. There are two commonly used sets of parameters resulting from treatments by Racah or by Condon and Shortley. Details of the calculations may be found for example in references 2, 3, 6, 7, or 9.

Three parameters are needed to describe interelectron repulsion between d-electrons. The Condon-Shortley parameters are labelled F_n and for d-electrons, F_0, F_2 and F_4 are required. The first of these, F_0, is a constant contribution to the energy of all terms in a given configuration (i.e. the coefficient of F_0 is the same for all the terms). The *separation* between two terms is thus a function of F_2 and F_4. The energies for the d^2 free ion terms are given in Table 2.1.

Table 2.3. *Racah Parameters for Transition Metal Free Ions.*
(The figures are B, C/B; the former in cm^{-1}).

| Element | Oxidation state | | | | |
	0	I	II	III	IV
Ti	560, 3·3	680, 3·7	720, 3·7		
Zr	250, 7·9	450, 3·9	540, 3·0		
Hf	280	440, 3·4			
V	580, 3·9	660, 4·2	765, 3·9	860, 4·8	
Nb	300, 8·0	260, 7·7	530, 3·8	600, 2·3	
Ta	350, 3·7	480, 3·8			
Cr	790, 3·2	710, 3·9	830, 4·1	1030, 3·7	1040, 4·1
Mo	460, 3·9	440, 4·5			680
W	370, 5·1				
Mn	720, 4·3	870, 3·8	960, 3·5	1140, 3·2	
Re	850, 1·4	470, 4·0			
Fe	805, 4·4	870, 4·2	1060, 4·1		
Ru	600, 5·4	670, 3·5	620, 6·5		
Co	780, 5·3	880, 4·4	1120, 3·9		
Ni	1025, 4·1	1040, 4·2	1080, 4·5		
Pd			830, 3·2		
Cu		1220, 4·0	1240, 3·8		

Ballhausen points out that the Racah formalism is the more complicated, but it has certain advantages for our purposes. Again three parameters A, B, and C are required. The relationships between A, B, C and F_0, F_2, and F_4 are [6]

$$A = F_0 - 49F_4,$$

$$B = F_2 - 5F_4,$$

$$C = 35F_4.$$

One advantage of using the Racah parameters is that the separation between the ground term and other terms of the same spin multiplicity is a function only of B, whereas it is a function of both F_2 and F_4. Thus the ^3F-^3P separation in d^2 (Table 2.1) is $15F_2-75F_4 = 15B$. The only excited terms we shall consider satisfy this criterion. A further advantage in calculations is that for a given configuration, the ratio C/B is approximately constant.

B (and C) are not constants, and in a complex, the value of B is always less than the free-ion value. This is discussed in Section 2.3.5. Values of B and C/B for some free ions are listed in Table 2.3.

2.2.2. SPIN–ORBIT COUPLING IN FREE IONS

The effects of spin–orbit coupling upon the free ion terms derived in Section 2.2.1 can be readily described as functions of the quantum numbers L and S of the terms.

The origin of spin–orbit coupling can be seen in a classical model as originating in the interaction of the electron spin with the magnetic field generated by the orbital motion of the electron. That is to say, l and s for a particular electron are no longer independent, and a new quantum number must be used to define the energy states. This quantum number, j, will take values of the vector sums of l and s.

The theory of spin–orbit coupling is rather complex. The reader is referred to reference 2 (pages 26–30) for a general discussion of the effect, or to reference 7 or 9 for a more complete treatment.

For a single electron, the spin–orbit coupling operator is

$$H' = \xi(r)\hat{l}.\,\hat{s}$$

where

$$\xi(r) = \frac{\hbar^2}{2m^2c^2}\frac{1}{r}\frac{\partial U(r)}{\partial r},$$

and the electron moves in a potential $U(r)$.

The energy of the electron in a wave-function defined by (n, l, j, m_j) is then

$$E(n, l, j, m_j) = E(n, l) + \zeta_{n,l} \int (n, l, j, m_j)^*\hat{l}.\,\hat{s}(n, l, j, m_j)\,d\tau$$

$$= E(n, l) + \tfrac{1}{2}(\zeta_{n,l})[j(j + 1) - l(l + 1) - s(s + 1)].$$

$\zeta_{n,l}$ is the single electron spin–orbit coupling constant (the symbol xi, ξ, is sometimes used for this constant also)

$$\zeta_{n,l} = \hbar^2 \int\limits_0^\infty [R(n, l)]^2 \xi(r) r^2 \, dr,$$

i.e. $\zeta_{n,l}$ is a radial integral, the usual device of parameterizing integrals involving the radial wave-function.

The two possible states of a single electron, $1 \pm \frac{1}{2}$, are then separated by $(1 + \frac{1}{2}) \zeta_{n,l}$.

If a Coulomb potential is assumed,

$$\zeta_{n,l} = \frac{e^2 \hbar^2}{2m^2 c^2 a_0^2} \frac{Z^4}{n^3 l(l + \frac{1}{2})(l + 1)}$$

where Z is the effective nuclear charge.

In the many electron case

$$H' = \sum_i \xi(r_i) \hat{l}_i \cdot \hat{s}_i.$$

The problem is then to transfer from this sum of single electron functions to an operator appropriate to a term specified by L and S.

Condon and Shortley [9] prove that the matrix element

$$E' = \int (L, M_L, S, M_S)^* H'(L, M'_L, S, M'_S) \, d\tau$$

must be proportional to an element

$$E' = \lambda \int (L, M_L, S, M_S)^* \hat{L} \cdot \hat{S}(L, M'_L, S, M'_S) \, d\tau,$$

where λ is the *term* spin–orbit coupling constant, i.e. for a given L and S, and is again a radial integral. There must be a connection between λ and $\zeta_{n,l}$, and it may be shown [2] that for a shell of electrons which is less than half-filled:

$$\lambda = \frac{\zeta_{n,l}}{2S},$$

while for a more than half-filled shell:

$$\lambda = -\frac{\zeta_{n,l}}{2S}.$$

It is clear from the expressions given above that $\zeta_{n,l}$ is a positive quantity for any realistic potential, $U(r)$. Thus λ changes sign depending on whether the shell is less, or more than half-filled. This is the origin of the inversion of spin–orbit coupling multiplets, which is referred to below. (When the shell *is* half-filled, $J = 0$ and $\lambda = 0$.)

Extrapolated values of $\zeta_{n,l}$ have been published by Dunn [10] for metal ions with 3d or 4d unpaired electrons, as a function of oxidation state. Some of these data are reproduced in Table 2.4.

Table 2.4. *Extrapolated values of the single electron spin–orbit coupling constants ζ_{3d} and ζ_{4d} for first and second row transition elements respectively as a function of oxidation states. Values are in cm^{-1}; those in parentheses are experimentally observed. (Reproduced from reference 10.)*

Oxidation state	M(0)	M(I)	M(II)	M(III)	M(IV)	M(V)	M(VI)
Ti	70	90	120	155			
V	(95)	135	170	210	250		
Cr	(135)	(190)	230	275	325	380	
Mn	(190)	255	(300)	355	415	475	540
Fe	255	345	400	(460)	515	555	665
Co	390	455	515	(580)	(650)	715	790
Ni		605	630	(715)	(790)	(865)	950
Cu			830	(875)	(960)	(1030)	(1130)
Zr	270	340	425	500			
Nb	(365)	490	555	670	750		
Mo	(450)	(630)	(695)	820	950	1030	
Tc	(550)	740	(850)	(990)	(1150)	(1260)	1450
Ru	745	900	1000	(1180)	(1350)	(1500)	(1700)
Rh	940	1060	1220	(1360)	(1570)	(1730)	(1950)
Pd		1420	1460	(1640)	(1830)	(2000)	(2230)
Ag			1840	(1930)	(2100)	(2300)	(2500)

It is conventional to refer to the energy levels resulting from interelectron repulsion as 'terms' and to those from spin–orbit coupling as 'states', however, the two descriptions are often used interchangeably.

The Russell-Saunders coupling scheme

Two possible schemes of coupling the angular momenta of sets of electrons must be considered. In the previous section we have tacitly assumed a Russell-Saunders coupling scheme. In this scheme, the strongest couplings are between the l_i of the electrons to give a new quantum number L and the s_i of the electrons to give a new quantum number S. L is the maximum value of Σm_l for a specific value of $S(=$ max. $\Sigma m_s)$. There is then a weaker spin–orbit coupling between L and S to give the new quantum number J where

$$J = L + S, L + S - 1 \ldots |L - S|,$$

a total of $2L + 1$ or $2S + 1$ values of J, whichever is least. Each state specified by J has a degeneracy of $2J + 1$ (i.e. there are M_J values $J, J - 1 \ldots -J$ in an external field). The separation between states is given by the Landé interval rule which is that the separation between *adjacent* states specified by J and $J + 1$ is $(J + 1)\lambda$. Thus the state with smallest J lies lowest when λ is positive, but the multiplet is inverted when λ is negative and the largest J lies lowest. These effects are illustrated in Fig. 2.1.

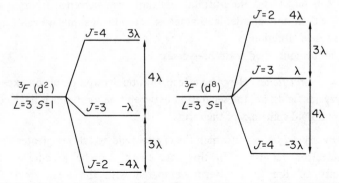

Figure 2.1. Splitting of the free ion ^3F term of d^2 and d^8 by spin–orbit coupling.

The Russell-Saunders coupling scheme is appropriate to the lighter elements, and is a good approximation at least for first row transition elements.

The j–j coupling scheme

In very heavy atoms, the Russell-Saunders coupling scheme is not suitable. In these cases, the strongest coupling is between l_i and s_i of a single electron to give a resultant j_i. The j's of individual electrons then couple less strongly to give J.

Pure j–j coupling is not observed, but an intermediate coupling scheme should be used for heavier elements.

2.3. Cubic crystal fields

We must now consider the effects of a crystal field upon the energies of free ion terms, and the effect of spin–orbit coupling upon any new energy levels. However, we must first establish the relative orders of magnitude of the effects of interelectron repulsion, crystal fields, and spin–orbit coupling. In this section only crystal fields of O_h or T_d symmetry will be considered.

2.3.1. THE STRENGTH OF CRYSTAL FIELDS

The effects of a crystal field upon the energy of an electron configuration are normally calculated using perturbation theory. As is shown in Chapter 3, this method depends upon the application of a small perturbation to a problem which is already solved. We must apply the largest perturbation first: if two perturbations produce similar changes in energy, they must be applied simultaneously. We are concerned with four perturbations of an electronic configuration: interelectron repulsion, crystal field, spin–orbit coupling, and magnetic field. The last of these always gives the smallest energy changes and so is considered last. When dealing with d-electrons, spin–orbit coupling is the smallest of the remaining three perturbations: this is not true for f-electrons because λ is larger and the f-orbitals are quite well screened from crystal field effects. We shall not consider lanthanide or actinide ions thus we can neglect this very weak field situation.

We shall be concerned with three cases:

(1) *Weak crystal fields.* Here the effects of interelectron repulsion exceed those due to crystal fields, i.e. the separation between free ion terms is large compared with crystal field splittings of the terms.

(2) *Strong crystal fields.* In which the crystal field effects are greater than those of interelectron repulsion. In this case there may be a reduction in spin multiplicity of the ground term compared with the weak field case, i.e. spin-pairing may take place. However, even in a configuration in which spin-pairing can occur, the strong field approximation may be the best even before spin-pairing takes place.

(3) *Intermediate crystal fields.* Where interelectron repulsion and the crystal field have comparable effects. A simultaneous perturbation treatment is now appropriate.

In each of the above three approximations, spin–orbit coupling and magnetic field perturbation are applied subsequently.

2.3.2. THE WEAK FIELD APPROXIMATION

Here we are interested in the effects of a crystal field upon the free ion terms discussed previously. The crystal field potential is an electrostatic potential which interacts with the orbital angular momentum of the electrons. Weak field terms have the same spin-degeneracy as their parent free ion term. Thus an S term will not be split by the crystal field because $L = 0$; also in O_h and T_d symmetry fields, P-terms are not split, analogous to the behaviour of p-electrons. Thus we need to consider D and F terms. Other excited terms occur, but the magnetochemist is mainly concerned with the ground free ion term.

(a) Fields of O_h symmetry. The wave-functions of the ^2D term of d^1 are the same combinations of single electron functions as the real forms of the d-orbitals, hence this term behaves like the d-orbitals to give two weak field terms $^2T_{2g}$ and ^2E_g. The separation between these terms is usually defined by the crystal field parameter Dq; $^2T_{2g}$ lies at $-4Dq$ and ^2E_g at $6Dq$, i.e. $10Dq$ apart (see Section 2.5 for definitions of D and q).

It should be remembered that the crystal field potential consists of two terms: one is spherically symmetrical and substantially raises the energy of all terms equally, the other is of O_h symmetry in this case, and leads to the splitting of the free ion term.

We use the group theory labels for these terms because the quantum number L is not a good quantum number in the crystal field, i.e. L cannot be used. This can also be seen because an E term has two-fold orbital degeneracy, which would require $L = \frac{1}{2}$. It is usual to label single electron wavefunctions (or orbitals) by lower case letters, e.g. d_{xy}, d_{yz} and d_{xz} are the t_{2g} orbitals, and crystal field terms by upper case letters, e.g. T_{2g}. A very common error (arising perhaps because d-orbitals and D terms have the same splitting pattern) is to try to place electrons in, e.g., the $^2T_{2g}$ level. This level is a consequence of pre-existing electronic arrangements and so it is wrong to argue that the $^2E_g \leftarrow ^2T_{2g}$ electronic transition necessarily represents the promotion of an electron from t_{2g} to e_g orbitals (the error is more serious, e.g. in d^2).

The orbital degeneracy of a term is one if the label is A or B; two if it is E; and three if it is T, irrespective of the subscript number (e.g. T_1 or T_2).

It will be seen from Table 2.2 that three other D ground terms occur: ^5D (d^4), ^5D (d^6), and ^2D (d^9). All three will give T_{2g} and E_g terms in an octahedral field, but we must decide which term will lie lowest.

^5D (d^6). A half-filled shell gives rise to a ^6S term and so is not split by the crystal field. It may thus be argued that d^6 is one electron more than a half-filled shell and so ^5D will behave like the ^2D of d^1. This the case and $^5T_{2g}$ lies lowest.

$^2D(d^9)$. This configuration is one electron *less* than the filled shell, the latter being also unaffected by the crystal field. It may thus be considered as one *positive* 'hole' in a filled shell and will be inverted relative to d^1 because of the formal reversal of charge. The ground term is therefore ^2E_g.

$^5D(d^4)$. This is treated as one positive 'hole' in a half-filled shell and so the lowest lying term is ^5E_g.

There is a formal, but perhaps not very helpful, resemblance between F-terms and f-orbitals (which give two triplets and a singlet in O_h fields). The ^3F term of d^2 gives three components in O_h: $^3T_{1g}$ at $-6Dq$, $^3T_{2g}$ at $2Dq$, and $^3A_{2g}$ at $12Dq$.

The order of the crystal field terms from the remaining d-configuration yielding F free ion terms can be obtained by arguing *via* filled or half-filled shell and the 'hole' formalism. These are given in Table 2.2. The P term does not split, but transforms as T_{1g} in the crystal field.

(b) *Crystal fields of T_d symmetry.* Just as the d-orbital splitting is inverted when comparing crystal fields of O_h or T_d symmetry, so are the splittings of free ion terms. Thus in T_d symmetry the ground terms are ^2E and $^3A_2$ for d^1 and d^2, respectively. The remainder are in Table 2.2.

The information in Table 2.2 should be part of the stock-in-trade of the magnetochemist, but it is easy to derive if the following is known: (i) the free ion terms for d^1 to d^5 since there is mirror symmetry about d^5; (ii) the splitting of ^2D (d^1) and ^3F (d^2) in O_h only, since all the rest can be obtained from half-filled or filled shell and 'hole' formalisms; and (iii) T_d is always inverted compared with O_h.

In the weak field approximation the overall splitting of a free ion term is small compared with the term separation, thus in d^2 $18Dq$ is small compared with $15B$.

2.3.3. STRONG CRYSTAL FIELDS

Discussion of the strong-field approximation is complicated by the possibility of spin-pairing, that is the lowest lying term may have a spin multiplicity less than the maximum possible, e.g. d^4 in O_h symmetry may have four unpaired electrons $(t_{2g}^3 e_g^1)$ and be 'high-spin' or only two unpaired electrons (t_{2g}^4) which is the 'low-spin' configuration. We must treat these cases separately, but can ignore the low-spin case in tetrahedral complexes because Dq is not large enough in these cases to bring about spin-pairing. (Dq for a tetrahedral complex is about $(-4/9)Dq$ for an octahedral complex with the same ligands.)

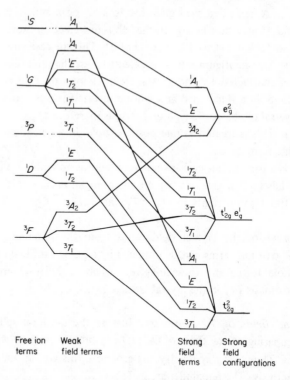

| Free ion terms | Weak field terms | Strong field terms | Strong field configurations |

Figure 2.2. Correlation diagram for the d^2 configuration in weak and strong octahedral crystal fields. Note that the lines joining weak and strong field terms do *not* represent the variation of their energies. The diagram is not to scale, and 'g' subscripts have been omitted from term labels.

(a) *High-spin complexes.* In this approximation, the effects of the crystal field dominate the energy thus the first step is to set up strong-field configurations defined by the number of electrons in the t_{2g} and e_g sets of orbitals (in O_h).

The case of the d^1 configuration is trivial since there is no interelectron repulsion. The ground strong-field configuration is t_{2g}^1 at $-4Dq$ and the only excited configuration is e_g^1 at $6Dq$. These correspond exactly to the $^2T_{2g}$ and ^2E_g weak field terms respectively and the energies of the two are linear in Dq at all values of Dq.

The d^2 configuration is different. Here there are three strong-field configurations: t_{2g}^2 at $-8Dq$, $t_{2g}^1e_g^1$ at $2Dq$ and e_g^2 at $12Dq$. These configurations will be split by interelectron repulsion to give strong-field terms. The derivation of the term labels from direct products is given, e.g. by Kettle [4]. The actual energies must be calculated in terms of Racah parameters. This situation is illustrated on the right-hand side of Fig. 2.2. This figure shows weak-field and strong-field terms for d^2.

Clearly these extremes must 'meet in the middle' and *qualitative* correlations are shown as straight lines. It will be noted that there are two $^3T_{1g}$ terms in each limit, the non-crossing rule requires us to correlate the two lower terms of the same degeneracy and symmetry *not* lower to higher and vice-versa. The origin of this rule must be investigated. Referring to Fig. 2.2, the ground term is always $^3T_{1g}$ but, ignoring inter-electron repulsion at the strong field limit, the energy of this term is $-6Dq$ in the weak field and $-8Dq$ in the strong field. A similar discrepancy exists for all four configurations with free ion F ground terms: it is closely related to the non-crossing rule. In the weak-field approximation, we only considered the first order effects of the crystal field upon the wave-functions of a particular free ion term. Thus in d^2 in a *weak* field the wave-functions of the lower $^3T_{1g}$ term are linear combinations of the ^3F term wave-functions only. In the strong-field approximation we must consider second-order terms which, if finite, will mix the wave-functions of different free ion terms. But, the criterion for such second-order terms to be finite is that the two sets of interacting functions must have the same symmetry and degeneracy. There are thus finite second-order interactions between the $^3T_{1g}(F)$ and $^3T_{1g}(P)$ terms originating from ^3F and ^3P, respectively. The interaction is always such as to lower the energy of the $^3T_{1g}(F)$ and raise that of $^3T_{1g}(P)$ by the same amount. Hence these two levels are further apart than the first order treatment predicts, moreover, the second-order term increases inversely as the first-order separation of the terms, whence the non-crossing rule since the second-order term becomes rapidly larger as the separation decreases.

We can discover the magnitude of this interaction by anticipating certain results of Chapter 3 and using the d^2 configuration as an example. It is necessary to calculate integrals using a Hamiltonian which we may write

$$\mathcal{H} = \mathcal{H}_0 + V_{oct} \tag{2.1}$$

where V_{oct} is the octahedral crystal field operator (see Section 2.5) and $\mathcal{H}_0$ is the Hamiltonian appropriate to a free ion. We shall use $|^3T^0_{1g}(F)\rangle$ and $|^3T^0_{1g}(P)\rangle$ to represent the wave-functions of the unmixed $^3T_{1g}(F)$ and $^3T_{1g}(P)$ terms. The bra-ket nomenclature is defined in Chapter 3.

Now, when $V_{oct} = 0$, the energy of $^3T_{1g}(F)$ is the same as that of ^3F, which we may take as zero; while that of $^3T_{1g}(P)$ will be $15B$ (i.e. the energy of ^3P). Thus

$$\left.\begin{array}{l} \langle^3T^0_{1g}(F)|\mathcal{H}_0|^3T^0_{1g}(F)\rangle = 0 \\[2mm] \text{and} \\[2mm] \langle^3T^0_{1g}(P)|\mathcal{H}_0|^3T^0_{1g}(P)\rangle = 15B \end{array}\right\} . \tag{2.2}$$

In the weak field, V_{oct} lowers the energy of $^3T_{1g}(F)$ to $-6Dq$ relative to ^3F but the $^3T_{1g}(P)$ term is unaffected. Thus

$$\left.\begin{array}{l} \langle^3T^0_{1g}(F)|V_{oct}|^3T^0_{1g}(F)\rangle = -6Dq \\[2mm] \text{and} \\[2mm] \langle^3T^0_{1g}(P)|V_{oct}|^3T^0_{1g}(P)\rangle = 0 \end{array}\right\} . \tag{2.3}$$

We need to evaluated matrix elements

$$\langle^3T^0_{1g}(F)|V_{oct}|^3T^0_{1g}(P)\rangle = x. \tag{2.4}$$

The secular determinant with $\mathcal{H}$ is

$$\begin{array}{c|cc} & ^3T^0_{1g}(F) & ^3T^0_{1g}(P) \\ \hline ^3T^0_{1g}(F) & -6Dq - E & x \\ ^3T^0_{1g}(P) & x & 15B - E \end{array} = 0, \tag{2.5}$$

i.e.

$$(-6Dq - E)(15B - E) - x^2 = 0. \tag{2.6}$$

We can evaluate (2.4) but it is more convenient to recognize that at the strong-field limit, not only may we ignore B, but (2.6) must give roots which are the energies of t^2_{2g} ($-8Dq$) and $t^1_{2g} e^1_g$ ($2Dq$) which are the strong-field configurations with which the two $^3T_{1g}$ terms correlate.

Ignoring B, (2.6) becomes

$$E^2 + 6DqE - x^2 = 0 \tag{2.7}$$

and (2.7) only has roots $-6Dq$ and $2Dq$ if

$$x = 4Dq.$$

Putting this value of x back in (2.6) gives us

$$E^2 + (6Dq - 15B)E - 16D^2q^2 - 90DqB = 0. \tag{2.8}$$

The roots of (2.8) are then the energies of the two $^3T_{1g}$ terms for all values of B and Dq. We have seen that (2.8) gives roots $-8Dq$ and $2Dq$ when $B = 0$; when $Dq = 0$ it becomes

$$E^2 - 15BE = 0$$

which has roots of 0 and $15B$, i.e. the free ion energies.

We shall also need the new wave-functions for the $^3T_{1g}(F)$ term which can be written as combinations

$$\Psi(^3T_{1g}(F)) = (1 + c_i^2)^{-\frac{1}{2}}[\Psi(^3T_{1g}^0(F)) + c_i\Psi(^3T_{1g}^0(P))]. \tag{2.9}$$

The process of obtaining c_i is to solve Equation 2.8 and substitute the lower root into the determinant (2.5) and solve for the coefficients of the wave-functions.

If the solutions to (2.8) are expressed as multiples of Dq, it is possible to rewrite it as

$$E^2 + (6 - 15B/Dq) - 16 - 90B/Dq = 0$$

or

$$E + (6 - Z) - 16 - 6Z = 0 \tag{2.10}$$

where

$$Z = \frac{15B}{Dq}.$$

Equation 2.10 is usually quoted in textbooks.

If E is the lower root of (2.10) it can be shown that

$$c_i = \frac{6Dq + E}{4Dq}. \tag{2.11}$$

Thus (2.9) and (2.11) give us the wave-functions for the $^3T_{1g}(F)$ term at any crystal field strength. At the weak-field limit $E = -6Dq$, thus $c_i = 0$, while at the strong-field limit $E = -8Dq$ whence $c_i = -\frac{1}{2}$.

It is convenient in magnetic calculations to introduce a new parameter A where

$$A = \frac{1\cdot5 - c_i^2}{1 + c_i^2}.$$

The reasons for this will be found in Chapter 4.

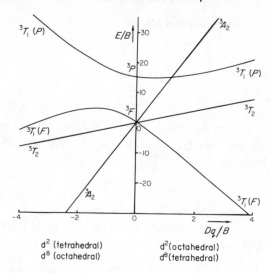

d² (tetrahedral)   d²(octahedral)
d⁸ (octahedral)   d⁸(tetrahedral)

Figure 2.3. Energies of the spin-triplet terms of d^2 and d^8 complexes as a function of crystal field strength.

Since (2.8) is a quadratic equation, the energies of the $^3T_{1g}$ terms will not be linear in Dq. Hence in Fig. 2.3 where the energies of the triplet levels of d^2 and d^8 are shown, the plots of the T_{1g} energies are curved, the others are linear since there is, for example, no other $^3T_{2g}$ term in the d^2 manifold.

Complete energy level diagrams (i.e. Tanabe-Sugano diagrams) will be found in any book on crystal field theory.

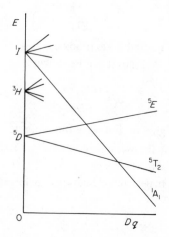

Figure 2.4. Qualitative variation of certain of the energy levels of an octahedral d^6 complex as a function of crystal field strength.

(b) *Low-spin complexes.* In such complexes, the separation between the t_{2g} and e_g orbitals has become sufficiently large that the increased interelectron repulsion resulting from pairing electrons in the t_{2g} orbitals is now less than the $t_{2g} \rightarrow e_g$ promotion energy.

Both the spin- and orbital-degeneracy may be reduced on spin-pairing. As an extreme example, the d^6 configuration has a $^5T_{2g}$ ground term in weak octahedral fields changing to a diamagnetic $^1A_{1g}$ strong-field ground term. A simplified energy level diagram for d^6 is given in Fig. 2.4: the ground term changes because the energy of the $^1A_{1g}$ component of a free ion ^1I term falls rapidly as Dq is increased, ultimately crossing the $^5T_{2g}$ weak-field ground term.

Ground terms for low-spin configurations are listed in Table 2.5.

Table 2.5. *Low-spin configurations. Ground strong-field configurations and ground strong field terms derived from them.*

Configuration	Ground strong field configuration	Ground strong field term in O_h symmetry
d^1	t_{2g}^1	$^2T_{2g}$ *
d^2	t_{2g}^2	$^3T_{1g}$ *
d^3	t_{2g}^3	$^4A_{2g}$ *
d^4	t_{2g}^4	$^3T_{1g}$
d^5	t_{2g}^5	$^2T_{2g}$
d^6	t_{2g}^6	$^1A_{1g}$
d^7	$t_{2g}^6 e_g^1$	^2E_g
d^8	$t_{2g}^6 e_g^2$	$^3A_{2g}$ *
d^9	$t_{2g}^6 e_g^3$	^2E_g *

* These are the same in the weak field limit.

2.3.4. INTERMEDIATE STRENGTH CRYSTAL FIELDS

In this region, a *qualitative* correlation between weak- and strong-field limits can be obtained as shown in Fig. 2.2. There is no option but to calculate the intermediate field region as simultaneous interelectron repulsion and crystal field perturbations. The results of these calculations are well known, usually being presented as the so-called Tanabe-Sugano diagrams [11]. We are fortunate in only wanting the energies of ground terms which have all been discussed above.

2.3.5. NEPHELAUXETIC AND SPECTROCHEMICAL SERIES

These series are of little direct interest in magnetochemistry except insofar as we may wish to estimate the value of c_i in Equation 2.11 which depends on B and Dq.

(a) *The nephelauxetic series.* It is observed that the value of B in a complex is always less than the appropriate free ion value. The extent to which B is reduced may be expressed in the nephelauxetic series. If we write $B_{obs.}/B_{free\ ion} = \eta$, the series is one showing increasing $(1 - \eta)$.

The interpretation of the series obtained is difficult. It must be related in part at least to covalent bonding. The series for ligands is:

$$F^- < H_2O < \text{urea} < NH_3 < \text{en} \simeq C_2O_4^{2-} < \underline{N}CS^- < Cl^- \simeq CN^- < Br^- < I^-.$$

(b) *The spectrochemical series.* These series simply place metal ions and ligands in order of increasing Dq. For metals (with a fixed ligand) Dq increases with increasing oxidation state, and in a triad for the same oxidation state increases from first to second to third transition series. Changes, e.g., within the first transition series, are more difficult to understand. The spectrochemical series for some ligands is:

$$I^- < Br^- < Cl^- \simeq \underline{S}CN^- < F^- \simeq \text{urea} \simeq OH^- < C_2O_4^{2-} < H_2O < \underline{N}CS^- < \text{pyridine}$$

$$\simeq NH_3 < \text{ethylenediamine} < \text{bipyridyl} \ll CN^-.$$

The corresponding series for metals is:

$$\text{Mn(II)} < \text{Ni(II)} < \text{Co(II)} < \text{Fe(II)} < \text{V(III)} < \text{Fe(III)} < \text{Cr(III)} < \text{Co(III)} <$$

$$\text{Mn(IV)} < \text{Mo(III)} < \text{Rh(III)} < \text{Pd(IV)} < \text{Ir(III)} < \text{Re(IV)} < \text{Pt(IV)}.$$

Jörgensen [12] has shown that both $(1 - \eta)$ and Dq may be factored into a contribution from the metal and one from the ligand. In the absence of actual spectroscopic data these are useful approximations.

2.4. Spin–orbit coupling in crystal field terms

We have already noted that L is not a good quantum number in the presence of a crystal field. It follows that we cannot describe the spin–orbit coupling states by a quantum number J. There is usually no alternative but to calculate the effect of spin–orbit coupling in each case since general rules cannot be formulated. In certain cases, second order spin–orbit coupling calculations will be important – both first and second order calculations are performed in detail in Chapter 4.

We list the values of all finite matrix elements with the operator $\lambda\hat{L}.\hat{S}$ in Chapter 4. It may perhaps be noted here that the matrix elements of spin–orbit coupling, $\langle\psi_i|\lambda\hat{L}.\hat{S}|\psi_j\rangle$ can only be finite when the M_L values of ψ_i and ψ_j differ by 0 or ±1.

The energies and degeneracies of the spin–orbit coupling states derived from T_{2g} and T_{1g} ground terms are also listed in Chapter 4.

2.5. Wave-functions and energy levels in cubic and axially distorted complexes

The octahedral crystal field operator, V_{oct}, and the crystal field parameters Dq

have been referred to previously. We have described the results, in terms of energies, of applying V_{oct} to give weak and strong field terms and have derived a form for the wave-functions of $^3T_{1g}(F)$ at all field strengths. However, we have not looked at the forms of the wave-functions of crystal field terms from D free ion terms or weak field terms derived from F terms. In this section we discuss the form of the operator V_{oct}. This operator may be defined taking either a four-fold or a three-fold axis of the octahedron as the z-axis: we shall define it with respect to both and similarly give both sets of wave-functions. The results, in terms of energies and symmetries of the crystal field terms must, of course, be the same irrespective of the choice of z-axis. We shall also discuss the effects of tetragonal and trigonal distortions of the octahedron on the energies and wave-functions of crystal field terms.

2.5.1. CUBIC SYMMETRY

Only the operator V_{oct} need be considered since the operator for tetrahedral fields differs only in sign and magnitude, giving splittings of $-\frac{4}{9}Dq$ relative to the octahedral case. V_{oct} can be expressed as a combination of certain of the spherical harmonics $Y_l^{m_l}(\theta,\phi)$. These are functions of the polar angles θ and ϕ, the numbers l and m_l taking the same values as the quantum numbers l and m_l. The spherical harmonics are listed, for example, in references 2 and 6. The same functions are used to represent the angular parts of electronic wave-functions, thus the angular parts of any p-electron functions are Y_1^0 and $Y_1^{\pm 1}$. Any of these functions which have $m_l > 0$ are imaginary, and to obtain the 'real' p-orbitals linear combinations must be taken thus:

$$p_z = Y_1^0,$$
$$p_x = \sqrt{\tfrac{1}{2}}(Y_1^{-1} - Y_1^1), p_y = i\sqrt{\tfrac{1}{2}}(Y_1^1 + Y_1^{-1}).$$

The form of V_{oct} can easily be deduced from group theory arguments, since the operator must be invariant under all operations of the group. The derivation is given by Ballhausen [2], while symmetry arguments are more fully applied by Bleaney and Stevens [13].

Choosing a four-fold axis as the z-axis,

$$V_{oct} = [Y_4^0 + \sqrt{\tfrac{5}{14}}(Y_4^4 + Y_4^{-4})]. \tag{2.12}$$

Similarly, referred to a three-fold axis,

$$V_{oct} = [Y_4^0 + \sqrt{\tfrac{10}{7}}(Y_4^3 + Y_4^{-3})]. \tag{2.13}$$

Neither (2.12) nor (2.13) contain imaginary parts. Both may be converted to Cartesian coordinates giving the same expression:

$$V_{oct} = D(x^4 + y^4 + z^4 - 3r^4/5). \tag{2.14}$$

The term D in Equation 2.14 is half of the crystal field parameter Dq. The second part, q, is introduced to circumvent the difficulty of calculating integrals involving the radial parts of the d-wave-functions. We need integrals $\langle m_l | V_{oct} | m_l' \rangle$: q is defined so that these integrals are simple multiples of Dq in the weak-field approximation. The energies and wave-functions in the weak-field approximation are listed in Table 2.6 for terms originating from free ion P, D

Table 2.6. *Energies and wave-functions of weak crystal field terms arising from free ion D and F ground terms and P terms with the same spin multiplicity as an F ground term. The wave-functions are referred to both a fourfold and a three-fold axis of an octahedron. Term energies are those of $^2D(d^1)$ and ^3F (d^2). Only the orbital part of the wave-function $|M_L\rangle$ is given: L is that appropriate to the parent free ion term and M_S values should be used as appropriate to the spin degeneracy of the parent term.*

Free ion term	Crystal field term	Energy	Wave-functions referred to Four-fold axis	Three-fold axis
D	T_{2g}	$-4Dq$	$\lvert 1\rangle$ $\lvert -1\rangle$ $\sqrt{\tfrac{1}{2}}(\lvert 2\rangle - \lvert -2\rangle)$	$\lvert 0\rangle$ $\sqrt{\tfrac{2}{3}}\lvert -2\rangle + \sqrt{\tfrac{1}{3}}\lvert 1\rangle$ $\sqrt{\tfrac{2}{3}}\lvert 2\rangle - \sqrt{\tfrac{1}{3}}\lvert -1\rangle$
	E_{g}	$6Dq$	$\lvert 0\rangle$ $\sqrt{\tfrac{1}{2}}(\lvert 2\rangle + \lvert -2\rangle)$	$\sqrt{\tfrac{1}{3}}\lvert -2\rangle - \sqrt{\tfrac{2}{3}}\lvert 1\rangle$ $\sqrt{\tfrac{1}{3}}\lvert 2\rangle + \sqrt{\tfrac{2}{3}}\lvert -1\rangle$
F	T_{1g}	$-6Dq$	$\lvert 0\rangle$ $\sqrt{\tfrac{3}{8}}\lvert -1\rangle + \sqrt{\tfrac{5}{8}}\lvert 3\rangle$ $\sqrt{\tfrac{3}{8}}\lvert 1\rangle + \sqrt{\tfrac{5}{8}}\lvert -3\rangle$	$\tfrac{2}{3}\lvert 0\rangle - \sqrt{\tfrac{5}{18}}(\lvert 3\rangle - \lvert -3\rangle)$ $\sqrt{\tfrac{5}{6}}\lvert 2\rangle + \sqrt{\tfrac{1}{6}}\lvert -1\rangle$ $\sqrt{\tfrac{5}{6}}\lvert -2\rangle - \sqrt{\tfrac{1}{6}}\lvert 1\rangle$
	T_{2g}	$2Dq$	$\sqrt{\tfrac{5}{8}}\lvert -1\rangle - \sqrt{\tfrac{3}{8}}\lvert 3\rangle$ $\sqrt{\tfrac{5}{8}}\lvert 1\rangle - \sqrt{\tfrac{3}{8}}\lvert -3\rangle$ $\sqrt{\tfrac{1}{2}}(\lvert 2\rangle + \lvert -2\rangle)$	$\sqrt{\tfrac{1}{2}}(\lvert 3\rangle + \lvert -3\rangle)$ $\sqrt{\tfrac{1}{6}}\lvert 2\rangle - \sqrt{\tfrac{5}{6}}\lvert -1\rangle$ $\sqrt{\tfrac{1}{6}}\lvert -2\rangle + \sqrt{\tfrac{5}{6}}\lvert 1\rangle$
	A_{2g}	$12Dq$	$\sqrt{\tfrac{1}{2}}(\lvert 2\rangle - \lvert -2\rangle)$	$\sqrt{\tfrac{2}{9}}(\lvert 3\rangle - \lvert -3\rangle) - \sqrt{\tfrac{5}{9}}\lvert 0\rangle$
P	T_{1g}		$\lvert 0\rangle$ $\lvert 1\rangle$ $\lvert -1\rangle$	$\lvert 0\rangle$ $\lvert 1\rangle$ $\lvert -1\rangle$

and F terms. Only the orbital parts are given: they are taken with the appropriate values of M_S for ^2D and ^5D terms and similarly for ^3F and ^4F terms. Wave-functions for ground T_{1g} terms from free ion F terms have already been given for stronger crystal fields in Equations 2.9 – 2.11. The wave-functions of terms originating from free ion D ground terms do not vary with crystal field strength. Wave-functions referred to both the four-fold and three-fold axes as z are listed in Table 2.6.

2.5.2. AXIALLY DISTORTED CUBIC SYMMETRIES

We are here concerned with the effect of distorting the crystal field along a four-fold or three-fold axis. Such a tetragonal or trigonal distortion may be treated by adding another term to the octahedral crystal field operator. In either case the additional term involves the spherical harmonic Y_2^0 thus

$$V = V_{oct} + Y_2^0. \tag{2.15}$$

It was noted in Section 2.5.1 that V_{oct} has the same Cartesian form referred to either axis and (2.15) can be written

$$V = D(x^4 + y^4 + z^4 - 3r^4/5) + C(z^2 - r^2/3). \tag{2.16}$$

In (2.16) C is a parameter similar to D in V_{oct}, and we can define a parameter, p (analogous to q), such that matrix elements

$$\langle m_l | z^2 - r^2/3 | m_l' \rangle$$

are simple multiples of p. There is then a parameter Cp which describes the new energies when cubic field terms are perturbed by an axial field component.

It is easy to see from group theory that splitting of the cubic field terms may arise. For example, if an octahedron is distorted by compression or extension along a four- or three-fold axis, the point group changes from O_h to D_{4h} or D_{3d} respectively. Neither of the latter groups contain triply degenerate representations, therefore all T terms of O_h may be expected to split in the distorted cases. The relevant parts of the correlation tables for O_h to D_{4h} and D_{3d} are given in Table 2.7.

Table 2.7. *Symmetry correlation tables for changes in point group from O_h to D_{4h} and D_{3d}.*

Representation in O_h	Representations in	
	D_{4h}	D_{3d}
A_{2g}	B_{1g}	A_{2g}
E_g	$A_{1g} + B_{1g}$	E_g
T_{1g}	$A_{2g} + E_g$	$A_{2g} + E_g$
T_{2g}	$B_{2g} + E_g$	$A_{1g} + E_g$

It should be noted that D_{4h} and D_{3d} are not the only possible point groups to arise from tetragonal or trigonal distortion of cubic complexes. Thus extension or compression of an octahedron along a three-fold axis reduces the symmetry to D_{3d} as does a distortion of a tetrahedron by compression of two opposite corners of a cube. If the three ligands off a three-fold axis of a tetrahedron are spread out, the symmetry will be C_{3v}. A tetrahedron may also be distorted to D_{2d} symmetry.

Distortion of an octahedron by rotating one triangular face about the three-fold axis so that the projection along that axis is no longer exactly staggered does not affect the degeneracy of the terms [14].

There is not necessarily a direct correlation between geometric distortion and the symmetry of the crystal field. It is the latter with which we are concerned in calculations whereas the former is experimentally accessible. Thus in an octahedral complex with six identical monodentate ligands it may be reasonable to equate geometric and crystal field symmetries. If on the other hand there are several different ligands the crystal field symmetry will depend both on their distances from the metal and their relative crystal field strengths. Contributions to the crystal field from counter ions may be significant in magnetism where different alkali metal salts of a single complex anion may have different magnetic properties.

It is possible to calculate the splittings of the cubic field terms as a function of Cp. These calculations are described by Figgis [14]. For the purposes of magnetic calculations, it is more convenient to use a parameter Δ to represent the separation of the components of a T term which result from either distortion. In both cases, the T_1 and T_2 terms split into orbital singlet and doublet: Δ is *defined* to be positive when the orbital singlet lies lowest. These results are used in Chapter 5. It can be seen from Table 2.8 that the sign of Cp depends on the symmetry of the axial field; it will also depend on the sense (i.e.

Table 2.8. *Wave-functions and energies of terms derived from cubic field T_2 and E terms in axially distorted crystal fields. The term energies are appropriate to d^1 or d^2. Wave-functions are those listed in Table 2.6 and may be taken with appropriate values of M_S.*

Term		Tetragonal field			Trigonal field		
Free ion	Cubic field	Term	Wave-function	Energy	Term	Wave-function	Energy
D	T_2	B_2	$\sqrt{\frac{1}{2}}(\lvert 2\rangle - \lvert -2\rangle)$	$-4Cp$	A_1	$\lvert 0\rangle$	$4Cp$
		E	$\lvert \pm 1\rangle$	$2Cp$	E	$\sqrt{\frac{2}{3}}\lvert \mp 2\rangle \pm \sqrt{\frac{1}{3}}\lvert \pm 1\rangle$	$-2Cp$
D	E	A_1	$\lvert 0\rangle$	$4Cp$	E	No splitting. Wave-functions as in Table 2.6	—
		B_1	$\sqrt{\frac{1}{2}}(\lvert 2\rangle + \lvert -2\rangle)$	$-4Cp$			
F	T_2	B_2	No splitting. Wave-functions as in Table 2.6		A_1	$\sqrt{\frac{1}{2}}(\lvert 3\rangle + \lvert -3\rangle)$	$-2Cp$
		E			E	$\sqrt{\frac{1}{6}}\lvert \pm 2\rangle \mp \sqrt{\frac{5}{6}}\lvert \mp 1\rangle$	Cp

compression or extension) of the distortion and on the d-configuration from which the term derives. The values of Cp in this Table are appropriate to d^1 and d^2 'octahedral' terms: the signs for other configurations can be obtained in the same ways as for cubic field terms (i.e. using half-filled shells and 'hole' formalisms).

The energies and wavefunctions for the T_2 and E cubic field terms arising from the free ion D term of d^1 and of the T_2 term from the free ion F term of d^2 are given in Table 2.8 for both tetragonal and trigonal distortions. In tetragonal cases the components of the latter T_2 term are accidentally degenerate.

The results for T_1 terms are more complicated because of the mixing between $T_1(F)$ and $T_1(P)$ terms in the crystal field which is discussed in Section 2.3.3. These wave-functions are defined by Equations 2.9–2.11.

The orbital parts of the wave-functions are, as $|L, M_L\rangle$,

$$(1 + c_i^2)^{-\frac{1}{2}}[|3, 0\rangle + c_i|1, 0\rangle] \equiv |0\rangle$$

and

$$(1 + c_i^2)^{-\frac{1}{2}}[\sqrt{\tfrac{15}{24}}|3 \pm 3\rangle + \sqrt{\tfrac{9}{24}}|3 \mp 1\rangle + c_i|1 \mp 1\rangle] \equiv |A\rangle.$$

The nomenclature $|0\rangle$ and $|A\rangle$ is explained in Chapter 4. The splitting, Δ, is then as given in Table 2.9.

Table 2.9. *Splitting, Δ, of the cubic field $T_{1g}(F)$ term from free ion F ground terms in axially distorted crystal fields. (Δ is positive when the orbital singlet lies lowest.)*

Distortion	Weak field $(c_i = 0)$	Immediate field	Strong field $(c_i = -\tfrac{1}{2})$
Tetragonal	$-2 \cdot 4Cp$	$[(42c_i^2 + 72c_i - 12)/(5 + 5c_i^2)]Cp$	$-6Cp$
Trigonal	$0 \cdot 6Cp$	$[(52c_i^2 - 48c_i + 3)/(5 + 5c_i^2)]Cp$	$6Cp$

In strong fields, the configuration t_{2g}^4 also gives rise to a $^3T_1$ term: the energies for this term are as in Table 2.9, but $c_i = -\tfrac{1}{2}$ must be used as this is a strong-field situation.

Concluding remarks

This very brief survey of crystal field theory completes the introductory part of the book. We now have those energies and wave-functions of crystal field terms which we shall need to proceed with detailed spin–orbit coupling and magnetic field calculations.

REFERENCES

1. D. Sutton (1968). *Electronic Spectra of Transition Metal Complexes*, McGraw-Hill, London.
2. C.J. Ballhausen (1962). *Introduction to Ligand Field Theory*, McGraw-Hill, New York.
3. B.N. Figgis (1966). *Introduction to Ligand Fields*, Interscience, New York.
4. S.F.A. Kettle (1969). *Coordination Compounds*, Nelson, London.
5. T.M. Dunn, D.S. McClure and R.G. Pearson (1965). *Some Aspects of Crystal Field Theory*, Harper and Row, New York.
6. H. Watanabe (1966). *Operator Methods in Ligand Field Theory*, Prentice-Hall, New Jersey.
7. J.S. Griffiths (1964). *The Theory of Transition-Metal Ions*, Cambridge University Press.
8. H. Bethe (1929). *Ann. Physik.*, **3**, 135.
9. E.U. Condon and G.H. Shortley (1963). *The Theory of Atomic Spectra*, Cambridge University Press.
10. T.M. Dunn (1961). *Trans. Farad. Soc.*, **57**, 1441.
11. Y. Tanabe and S. Sugano (1954). *J. Phys. Soc. Japan*, **9**, 753.
12. C.K. Jörgensen (1962). *Absorption Spectra and Chemical Bonding in Complexes*, Pergamon, London.
13. B. Bleaney and K.W.H. Stevens (1953). *Reports Prog. Physics*, **16**, 126.
14. B.N. Figgis (1965). *J. Chem. Soc.*, 4887.

3 | Perturbation theory and its applications

In order that the reader may gain the maximum benefit from the subsequent chapters it is desirable to acquire an understanding of the quantum mechanical methods used in deriving the expressions for magnetic susceptibilities. Such an understanding is essential for the reader who ultimately wishes to do research in the subject and may have to perform calculations on systems which have not yet been studied. At this stage, having mentioned that quantum mechanics is involved, many chemists become discouraged and turn away from the subject. However, our own experience is that in most cases there are two main hurdles to overcome. The first one is that the mathematics does look complicated, but the amount required in order to calculate magnetic susceptibilities, and also electron spin resonance parameters, is no more than that contained in many university chemistry courses. The second hurdle is essentially that of a language barrier in that most quantum mechanical texts use a terminology unfamiliar to the average chemist, often without adequate explanation.

The aim of this chapter is to attempt to overcome the above difficulties by *outlining* the derivation of the quantum mechanical results we wish to use, explaining in detail what these results mean and also giving simple explanations where possible of the terminology being used. Having reached this stage the more interested reader is advised to enquire more deeply into the quantum mechanics involved.

3.1. Perturbation theory [1, 2]

The determination from first principles of the energy levels for a system such as a transition metal complex is extremely difficult. It requires the solution of the equation

$$\mathcal{H}\psi = E\psi \tag{3.1}$$

where the Hamiltonian operator $\mathcal{H}$ contains a large number of terms. For example in the case of a transition metal complex in a magnetic field we may have

$$\mathcal{H} = -\sum_i \frac{h^2}{8\pi^2 m} \nabla_i^2 - \sum_i \frac{Ze^2}{r_i} + \sum_{j>i} \frac{e^2}{r_{ij}} + V_L + \sum_i \zeta \hat{l}_i \cdot \hat{s}_i + \hat{\mu}H. \tag{3.2}$$

Reading from left to right the terms in the Hamiltonian represent:

(1) the kinetic energy of the electrons,
(2) the potential energy due to the attraction between the nucleus and electrons,
(3) the interelectronic repulsions,
(4) the potential due to the crystal field,
(5) the spin–orbit coupling,
(6) the interaction of the electrons with the applied magnetic field.

The usual way of tackling a problem of this complexity is to use an approximate method, the most useful of which for our purposes is perturbation theory. In this method the problem we wish to solve is considered from the point of view of a small perturbation on a problem which has already been solved. For example a transition metal ion in an octahedral crystal field would be treated as the potential due to the ligands being a perturbation on the free ion. Thus the treatment would start with the free ion energy levels and wave-functions, and then calculate the changes in the energies and wave-functions caused by the crystal field.

In this chapter we will outline the essential parts of perturbation theory (for a more detailed treatment the reader is referred to quantum mechanical texts such as *Quantum Chemistry* by H. Eyring, J. Walter and G. E. Kimball, Wiley (1964), or *Introduction to Quantum Mechanics*, L. Pauling and E. B. Wilson, McGraw-Hill (1934)). At this stage, where the discussion is in general terms, we hope that the reader meeting the subject for the first time will not be disheartened by the mathematics involved. As much descriptive explanation as possible is included in this chapter, and a number of examples of the application of perturbation theory to actual calculations is given in Chapters 4 and 5.

The problem we wish to solve will be assumed to have a Hamiltonian $\mathcal{H}$ only slightly different from that of a problem which has already been solved, i.e.

$$\mathcal{H} = \mathcal{H}_0 + \alpha \mathcal{H}^{(1)} \tag{3.3}$$

where α is a parameter,* and $\alpha \mathcal{H}^{(1)}$ is the perturbation which is small compared to $\mathcal{H}_0$.

For the problem which has already been solved we have

$$\mathcal{H}_0 \psi_n^0 = E_n^0 \psi_n^0 \tag{3.4}$$

where the energies (eigenvalues) are $E_1^0, E_2^0, \ldots E_n^0$ with wave-functions (eigenfunctions) $\psi_1^0, \psi_2^0, \ldots \psi_n^0$ corresponding to these energies. The equation we now wish to solve is

$$(\mathcal{H}_0 + \alpha \mathcal{H}^{(1)})\psi_n = E_n \psi_n. \tag{3.5}$$

If $\alpha = 0$, Equation 3.5 reduces to Equation 3.4, and we therefore expect that for small values of α the solutions to Equation 3.5 will be near to those of Equation 3.4. In other words the effects of the perturbation is to change slightly the unperturbed energies and wave-functions.

The method of solving Equation 3.5 depends on whether or not the unperturbed energy levels are degenerate. An outline of the solution of this equation for the non-degenerate case will be given first, since it is the simplest to deal with whilst remaining instructive. However, for the purposes of calculating magnetic susceptibilities the results are of little value since in nearly all of the cases in which we are interested the unperturbed energy levels (i.e. those before the application of a magnetic field) will at least have spin degeneracy.

3.1.1. THE NON-DEGENERATE CASE

The first step in the solution is to expand the wave-functions and energies as power series in α.

$$\psi_n = \psi_n^0 + \alpha \psi_n^{(1)} + \alpha^2 \psi_n^{(2)} + \ldots \tag{3.6}$$

$$E_n = E_n^0 + \alpha E_n^{(1)} + \alpha^2 E_n^{(2)} + \ldots \tag{3.7}$$

where $\psi_n^{(m)}$ and $E_n^{(m)}$ are independent of α and have to be determined. Substituting these series for ψ_n and E_n into Equation 3.5 and equating the coefficients of like powers of α, gives a series of equations:

$$\mathcal{H}_0 \psi_n^0 = E_n^0 \psi_n^0, \tag{3.8}$$

$$(\mathcal{H}_0 - E_n^0)\psi_n^{(1)} = E_n^{(1)}\psi_n^0 - \mathcal{H}^{(1)}\psi_n^0, \tag{3.9}$$

$$(\mathcal{H}_0 - E_n^0)\psi_n^{(2)} = E_n^{(2)}\psi_n^0 + E_n^{(1)}\psi_n^{(1)} - \mathcal{H}^{(1)}\psi_n^{(1)}. \tag{3.10}$$

* We use the parameter α rather than λ (used by H. Eyring, J. Walter and G.E. Kimball, and by L. Pauling and E.B. Wilson) in order to avoid any possible confusion with the spin–orbit coupling constant λ which has already been used.

Equation 3.8 represents the unperturbed situation for which we have solutions, whilst solving Equations 3.9 and 3.10 gives the unknowns $\psi_n^{(1)}$, $\psi_n^{(2)}$, $E_n^{(1)}$ and $E_n^{(2)}$. The solution of Equation 3.9 is obtained by expanding $\psi_n^{(1)}$ and $\mathcal{H}^{(1)}\psi_n^0$ in terms of a set of normalized and orthogonal functions ψ_1^0, ψ_2^0. . . ., ψ_n^0† i.e.

$$\psi_n^{(1)} = A_1\psi_1^0 + A_2\psi_2^0 + \ldots + A_m\psi_m^0 + \ldots \tag{3.11}$$

and

$$\mathcal{H}^{(1)}\psi_n^0 = I_{1n}^{(1)}\psi_1^0 + I_{2n}^{(1)}\psi_2^0 + \ldots + I_{mn}^{(1)}\psi_m^0 + \ldots \tag{3.12}$$

where

$$I_{mn}^{(1)} = \int \psi_m^{0*}\mathcal{H}^{(1)}\psi_n^0 \, d\tau.$$

In practice the functions ψ_1^0, ψ_2^0, . . . ψ_m^0 . . . will be the set of wave-functions which describe the unperturbed system. Substituting the series (3.11) and (3.12) into Equation 3.9 and equating the coefficients of each term in ψ_n^0 leads to the following expressions for the energy and wave-function of the nth level, to the first order in α:

$$E_n = E_n^0 + \alpha I_{nn}^{(1)} \tag{3.13}$$

$$\psi_n = \psi_n^0 + \alpha \sum_m{}' \frac{I_{mn}^{(1)}}{E_n^0 - E_m^0} \psi_m^0 \tag{3.14}$$

where Σ_m' means that the sum is over all values of m except n.

The interpretation of Equations 3.13 and 3.14 is that to a first order approximation the energy of the perturbed state is that of the unperturbed state plus an amount equal to the integral (or matrix element) $\int\psi_n^{0*}\mathcal{H}^{(1)}\psi_n^0 \, d\tau$, see Fig. 3.1.

At the same time the wave-function describing the perturbed state is that of the unperturbed state plus admixtures of wave-functions which were not degenerate with it in the unperturbed state. This modification of the wave-function under the action of the perturbation is commonly referred to as *mixing-in* due to the perturbation. The extent of the mixing between any pair of wave-functions depends on the value of the matrix element $I_{mn}^{(1)} = \int\psi_m^0 *\mathcal{H}^{(1)}\psi_n^0 d\tau$ and inversely on the energy separation between them.

The calculation may be taken to the second order in α, by a similar process to that already outlined. We will find that very often we wish to know the energy of a given state to this degree of approximation, although rarely do we wish to

† For further details of the expansion of functions see *Introduction to Quantum Mechanics*, L. Pauling and E.B. Wilson, Chapter 6, and *Quantum Chemistry*, H. Eyring, J. Walter, and G.E. Kimball, Chapter 3 page 31.

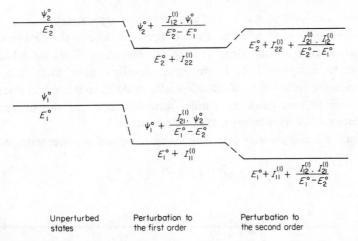

| Unperturbed states | Perturbation to the first order | Perturbation to the second order |

Figure 3.1. An illustration of the changes in energies and wave-functions of two non-degenerate levels when a perturbation is considered to the first and second order of approximation.

know the wave-function as accurately. The expression for the energy to the second order is:

$$E_n = E_n^0 + \alpha I_{nn}^{(1)} + \alpha^2 \sum_m' \frac{I_{nm}^{(1)} I_{mn}^{(1)}}{E_n^0 - E_m^0}. \tag{3.15}$$

In practice the parameter α is incorporated into $\mathcal{H}^{(1)}$ and hence it becomes unity in Equation 3.15.

3.1.2. THE DEGENERATE CASE

All the problems we wish to discuss in this book start with the unperturbed energy levels being degenerate and the expressions we have previously given are no longer applicable. If we apply a perturbation to an n-fold degenerate level we need to solve the equation:

$$(\mathcal{H}_0 + \alpha \mathcal{H}^{(1)})\psi = E\psi. \tag{3.16}$$

As $\alpha \to 0$ the energy E and wave-function must approach the solutions to the n-fold degenerate situation represented by:

$$\mathcal{H}_0 \psi^0 = E^0 \psi^0. \tag{3.17}$$

Assuming that the n linearly independent orthogonal wave-functions in the n-fold degenerate level are $\psi_1^0, \psi_2^0, \psi_3^0 \ldots \psi_n^0$ and the energy is $E_1^0 = E_2^0 \ldots = E_n^0$, then ψ must approach a solution to Equation 3.17; that is some linear combination of the ψ_i^0:

$$\phi = c_1 \psi_1^0 + c_2 \psi_2^0 + \ldots + c_n \psi_n^0. \tag{3.18}$$

The linear combination ϕ is known as a *zero order approximation to* ψ. These linear combinations are called zero order approximations to the wave-functions after perturbation because they are the combinations of the ψ_i^0 which are solutions to Equation 3.16. If we could somehow guess these zero order wave-functions the solution of Equation 3.16 would be facilitated. In practice it is not easy to guess these zero order wave-functions, and the first step in degenerate perturbation theory is their determination.

As for the non-degenerate case, E and ψ are expressed as power series in α:

$$E = E_1^0 + \alpha E^{(1)} + \alpha^2 E^{(2)} + \ldots, \tag{3.19}$$

$$\psi = \left(\sum_{j=1}^{n} c_j \psi_j^0 \right) + \alpha \psi^{(1)} + \alpha^2 \psi^{(2)} + \ldots . \tag{3.20}$$

Substituting these expansions in Equation 3.16, and then equating coefficients of like powers of α gives a series of equations involving the unknowns c_j, $\psi^{(1)}$, $\psi^{(2)}$, $E^{(1)}$, and $E^{(2)}$. For example to the zero and first order in α we have:

$$\mathcal{H}_0 \sum_{j=1}^{n} c_j \psi_j^0 = E_1^0 \sum_{j=1}^{n} c_j \psi_j^0, \tag{3.21}$$

$$(\mathcal{H}_0 - E_1^0)\psi^{(1)} = \sum_{j=1}^{n} c_j(E^{(1)} - \mathcal{H}^{(1)})\psi_j^0. \tag{3.22}$$

Equation 3.21 is already satisfied since this represents the unperturbed situation. As in the non-degenerate case Equation 3.22 is solved by putting

$$\psi^{(1)} = \sum A_j \psi_j^0 \quad \text{and} \quad \mathcal{H}^{(1)} \psi_j^0 = \sum_k I_{kj}^{(1)} \psi_k^0,$$

where $I_{kj}^{(1)} = \int \psi_k^{0*} \mathcal{H}^{(1)} \psi_j^0 d\tau$ and the A's are constants to be determined. Substituting these expansions into Equation 3.22 and equating the coefficients of ψ_j^0 enables the unknowns $E^{(1)}$, $\psi^{(1)}$, and c_j to be determined. This process results in the following set of the n simultaneous equations involving the c_j's, $E^{(1)}$, and $I_{kj}^{(1)}$.

$$(I_{11}^{(1)} - E^{(1)})c_1 + I_{12}^{(1)}c_2 + \ldots + I_{1n}^{(1)}c_n = 0,$$
$$I_{21}^{(1)}c_1 + (I_{22}^{(1)} - E^{(1)})c_2 + \ldots + I_{2n}^{(1)}c_n = 0, \tag{3.23}$$
$$I_{n1}^{(1)}c_1 + I_{n2}^{(1)}c_2 + \ldots + (I_{nn}^{(1)} - E^{(1)}c_n) = 0.$$

One possible solution to this set of equations is the trivial one where $c_1 = c_2 = \ldots c_n = 0$, and this is the only solution unless the determinant of the coefficients of the c's is equal to zero; that is unless:

$$\begin{vmatrix} (I_{11}^{(1)} - E^{(1)}) & I_{12}^{(1)} & \cdots & I_{1n}^{(1)} \\ I_{21}^{(1)} & (I_{22}^{(1)} - E^{(1)}) & \cdots & I_{2n}^{(1)} \\ \cdot & \cdot & & \cdot \\ \cdot & \cdot & & \cdot \\ \cdot & \cdot & & \cdot \\ I_{n1}^{(1)} & I_{n2}^{(1)} & & (I_{nn}^{(1)} - E^{(1)}) \end{vmatrix} = 0 \qquad (3.24)$$

This determinant is known as the *secular determinant*. Since the matrix elements $I_{nm}^{(1)}$ can usually be evaluated, this determinant is an equation of the nth degree in $E^{(1)}$ and therefore has n roots, $E_1^{(1)}, E_2^{(1)}, \ldots, E_n^{(1)}$ some of which may be equal. The zeroth order wave-functions corresponding to the lth root, $E_l^{(1)}$, is found by substituting this value of $E_l^{(1)}$ in the set of simultaneous Equations 3.23 and solving for the ratios $c_2/c_1, c_3/c_1, \ldots c_n/c_1$. These ratios plus the normalizing condition

$$c_1^* c_1 + c_2^* c_2 + \ldots + c_n^* c_n = 1$$

determine the c's and thus the zeroth order wave-function ϕ_l. It is interesting to note that in determining the zeroth order wave-functions we have also obtained the first order corrections to the energies. The process of finding the zeroth order combinations of the wave-functions and the first order energies from the secular determinant is often referred to as *diagonalization* of the determinant. This terminology arises because if we apply the perturbation to our set of zeroth order wave-functions we obtain a diagonal determinant, i.e. the matrix elements $I_{mn}^{(1)} = 0$ unless $m = n$, and the first-order corrections to the energies are simply $E^{(1)} = I_{nn}^{(1)}$.

Comparison of the coefficients of α^2, and working through the algebra as already outlined, results in the following expressions for the wave-function of the lth level to the first order and for the energy to the second order:

$$\psi_l = \phi_l + \alpha \sum_{j > n} \frac{\sum_{k=1}^{n} c_k I_{jk}^{(1)} \cdot \psi_j^0}{(E_1^0 - E_j^0)}. \qquad (3.25)$$

$$E_l = E_1^0 + \alpha E_l^{(1)} + \alpha^2 \sum_{j > n} \frac{I_{lj}^{(1)} I_{jl}^{(1)}}{(E_1^0 - E_j^0)}. \qquad (3.26)$$

As before α is normally incorporated in $\mathcal{H}^{(1)}$ and thus becomes unity in Equations 3.25 and 3.26.

The interpretation of Equation 3.25 is that to a first-order approximation the wave-function of the lth level after perturbation consists of a linear combination of the originally degenerate wave-functions plus some admixture of wave-functions which were not contained within the originally degenerate level. This point is illustrated with a simple example in Fig. 3.2. The amount of admixture is inversely proportional to the difference in energy between the states being mixed. The second-order correction to the energy shows a similar dependence, see Equation 3.26.

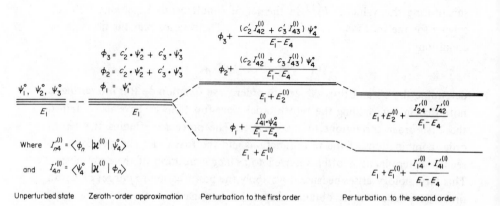

Figure 3.2. An illustration of the possible effects of a perturbation on a triply degenerate ground state and a singly degenerate excited state.

The details of the application of perturbation theory to some specific examples will be delayed until Chapter 4, where the calculation of magnetic susceptibilities is discussed in detail.

3.2. Some remarks on the evaluation of matrix elements

The solution of the secular determinant and the amount of mixing of various wave-functions due to a given perturbation, depend upon being able to evaluate matrix elements of the type $\int \psi_m^* \mathscr{H}^{(1)} \psi_n \, d\tau$. A fairly simple discussion of this will now be given using some of the perturbing operators and wave-functions of the type used in subsequent chapters. In many texts on quantum mechanics, and in the remainder of this work, a somewhat simpler notation is used to represent matrix elements of the above type. This notation is known as 'bra-ket' notation with the correspondence as follows:

$$\langle \psi_m | \mathscr{H} | \psi_n \rangle \equiv \int \psi_m^* \mathscr{H} \psi_n \, d\tau, \qquad (3.27)$$

i.e. $|\psi_n\rangle$ represents the wave-function ψ_n whilst $\langle\psi_n|$ represents its complex conjugate.

The steps in evaluating a matrix element of the type $\langle\psi_m|\mathscr{H}|\psi_n\rangle$ are:

(i) operate with $\mathscr{H}$ on the wave-function on its right-hand side, i.e. evaluate $\mathscr{H}|\psi_n\rangle$. This will result in a constant multiplied by a wave-function which may or may not be the same as the original. For the present let us assume $\mathscr{H}|\psi_n\rangle = c_i|\psi_n\rangle$.

(ii) The result of (i) is then premultiplied by $\langle\psi_m|$ giving $\langle\psi_m|c_i\psi_n\rangle$.

(iii) since c_i is a constant we have $\langle\psi_m|c_i\psi_n\rangle = c_i\langle\psi_m|\psi_n\rangle$ and we are thus left with the task of evaluating $\langle\psi_m|\psi_n\rangle$. Provided $|\psi_m\rangle$ and $|\psi_n\rangle$ are ortho-normalized, $\langle\psi_m|\psi_n\rangle = 1$ when $m = n$ but is zero otherwise.

The main perturbations with which we will be concerned in the present work are those of spin–orbit coupling and a magnetic field. The perturbations are represented by the energy operators $\lambda\hat{L}.\hat{S}$ and $H.(\hat{L} + 2\hat{S})$ respectively. (This latter operator should be $\beta H(\hat{L} + 2{\cdot}0023\hat{S})$, but unless otherwise stated we will assume that this approximation is a good one). All symbols with a circumflex denote *operators*, whilst symbols such as L, S, M_L, M_S are *quantum numbers* and are used to specify wave-functions. Both of the perturbations we have mentioned have been represented in terms of angular momentum operators. The spin–orbit coupling operator $\lambda\hat{L}.\hat{S}$ may be expanded in terms of the x, y, and z components of the angular momentum operators:

$$\lambda\hat{L}.\hat{S} = \lambda(\hat{L}_x\hat{S}_x + \hat{L}_y\hat{S}_y + \hat{L}_z\hat{S}_z) \tag{3.27}$$

Also we are usually interested in the effect of the magnetic field when it is applied along the principal axes of the molecule in which case the components of the magnetic moment operator are:

$$\hat{\mu}_z = \beta H_z(\hat{L}_z + 2\hat{S}_z),$$
$$\hat{\mu}_x = \beta H_x(\hat{L}_x + 2\hat{S}_x),$$
$$\hat{\mu}_y = \beta H_y(\hat{L}_y + 2\hat{S}_y). \tag{3.28}$$

A simple set of rules are available for the evaluation of the effect of these angular momentum operators on wave-functions. If $\hat{L}_x$, $\hat{L}_y$, $\hat{S}_x$ and $\hat{S}_y$ are expressed in terms of the shift operators $\hat{L}_+$, $\hat{L}_-$, $\hat{S}_+$, and $\hat{S}_-$ i.e.

$$\hat{L}_x = \tfrac{1}{2}(\hat{L}_+ + \hat{L}_-), \hat{S}_x = \tfrac{1}{2}(\hat{S}_+ + \hat{S}_-)$$
$$\hat{L}_y = \frac{-i}{2}(\hat{L}_+ - \hat{L}_-), \hat{S}_y = \frac{-i}{2}(\hat{S}_+ - \hat{S}_-), \tag{3.29}$$

then the rules are:

$$\hat{L}_z|L, M_L, S, M_S\rangle = M_L|L, M_L, S, M_S\rangle \tag{1}$$

$$\hat{L}_+|L, M_L, S, M_S\rangle = [L(L + 1) - M_L(M_L + 1)]^{\frac{1}{2}}|L, M_L + 1, S, M_S\rangle \tag{2}$$

$$\hat{L}_-|L, M_L, S, M_S\rangle = [L(L + 1) - M_L(M_L - 1)]^{\frac{1}{2}}|L, M_L - 1, S, M_S\rangle \tag{3}$$

$$\hat{S}_z|L, M_L, S, M_S\rangle = M_S|L, M_L, S, M_S\rangle \tag{4}$$

$$\hat{S}_+|L, M_L, S, M_S\rangle = [S(S + 1) - M_S(M_S + 1)]^{\frac{1}{2}}|L, M_L, S, M_S + 1\rangle \tag{5}$$

$$\hat{S}_-|L, M_L, S, M_S\rangle = [S(S + 1) - M_S(M_S - 1)]^{\frac{1}{2}}|L, M_L, S, M_S - 1\rangle \tag{6}$$

$$(3.30)$$

Equation 3.30(1) means that the operator $\hat{L}_z$ operates on the part of the wave-function specified by the quantum number M_L and leaves the wave-function unchanged but multiplies it by the *value* of M_L. The consequence of this is that the only non-zero matrix elements, $\langle\psi_m|\hat{L}_z|\psi_n\rangle$, are those for which ψ_m and ψ_n are identical. On the other hand, if the matrix element $\langle\psi_m|\hat{L}_+|\psi_n\rangle$, is to be non-zero then the only difference between ψ_m and ψ_n must be that ψ_m has an M_L value one unit greater than ψ_n.

Example 1

Evaluate the matrix element

$$I = \langle L, M_L + 1, S, M_S|\hat{L}_x|L, M_L, S, M_S\rangle.$$

Using Equation 3.29 to express $\hat{L}_x$ in terms of the shift operators we may write

$$I = \langle L, M_L + 1, S, M_S|\tfrac{1}{2}(\hat{L}_+ + \hat{L}_-)|L, M_L, S, M_S\rangle$$

$$= \tfrac{1}{2}\langle L, M_L + 1, S, M_S|\hat{L}_+|L, M_L, S, M_S\rangle$$

$$+ \tfrac{1}{2}\langle L, M_L + 1, S, M_S|\hat{L}_-|L, M_L, S, M_S\rangle$$

$$= \tfrac{1}{2}\langle L, M_L + 1, S, M_S|[L(L + 1) - M_L(M_L + 1)]^{\frac{1}{2}}|L, M_L + 1, S, M_S\rangle$$

$$+ \tfrac{1}{2}\langle L, M_L + 1, S, M_S|[L(L + 1) - M_L(M_L - 1)]^{\frac{1}{2}}|L, M_L - 1, S, M_S\rangle$$

$$= \tfrac{1}{2}[L(L + 1) - M_L(M_L + 1)]^{\frac{1}{2}}\langle L, M_L + 1, S, M_S|L, M_L + 1, S, M_S\rangle$$

$$+ \tfrac{1}{2}[L(L + 1) - M_L(M_L - 1)]^{\frac{1}{2}}\langle L, M_L + 1, S, M_S|L, M_L - 1, S, M_S\rangle$$

$$= \tfrac{1}{2}[L(L + 1) - M_L(M_L + 1)]^{\frac{1}{2}}$$

since

$$\langle L, M_L + 1, S, M_S| L, M_L + 1, S, M_s\rangle = 1$$

and

$$\langle L, M_L + 1, S, M_S| L, M_L - 1, S, M_S\rangle = 0.$$

Example 2

Evaluate the matrix element

$$I = \langle L, M_L, S, M_S | \hat{L}_z \hat{S}_z | L, M_L, S, M_S \rangle.$$

Using Equations 3.30(1) and 3.30(4) and proceeding in a stepwise manner

$$I = \langle L, M_L, S, M_S | \hat{L}_z M_S | L, M_L, S, M_S \rangle$$

$$= \langle L, M_L, S, M_S | M_L M_S | L, M_L, S, M_S \rangle$$

$$= M_L M_S$$

since

$$\langle L, M_L, S, M_S | L, M_L, S, M_S \rangle = 1$$

3.3. Orders of perturbation. First- and second-order Zeeman effects

In Chapter 1 the energy of a given level in a magnetic field was expressed as a power series in the magnetic field.

$$W_i = W_i^0 + W_i^{(1)}H + W_i^{(2)}H^2 + \ldots \tag{3.31}$$

The term $W_i^{(1)}H$ represents the change in energy to the first order in the magnetic field and is known as the first-order Zeeman effect. The change in energy to the second order in the magnetic field is given by the term $W_i^{(2)}H^2$ and is known as the second-order Zeeman effect. The first- and second-order Zeeman coefficients $W_i^{(1)}$ and $W_i^{(2)}$ may readily be evaluated using perturbation theory. If we consider for simplicity the hypothetical situation of a non-degenerate state ψ_n at energy W_n^0 perturbed by a magnetic field in the z-direction, then the energy will be

$$W_n = W_n^0 + W_n^{(1)}H_z + W_n^{(2)}H_z^2 + \ldots \tag{3.32}$$

However, non-degenerate perturbation theory (Equation 3.15) gives the following expression for the energy of nth level in the magnetic field:

$$W_n = W_n^0 + I_{nn}^{(1)} + \sum_m' \frac{I_{nm}^{(1)} I_{mn}^{(1)}}{W_n^0 - W_m^0}$$

$$= W_n^0 + \langle \psi_n | \beta H_z (\hat{L}_z + 2\hat{S}_z) | \psi_n \rangle$$

$$+ \sum_m' \frac{\langle \psi_n | \beta H_z (\hat{L}_z + 2\hat{S}_z) | \psi_m \rangle \langle \psi_m | \beta H_z (\hat{L}_z + 2\hat{S}_z) | \psi_n \rangle}{W_n^0 - W_m^0}$$

$$= W_n^0 + \beta H_z \langle \psi_n | \hat{L}_z + 2\hat{S}_z | \psi_n \rangle$$

$$+ \sum_m' \frac{\beta^2 H_z^2 \langle \psi_n | \hat{L}_z + 2\hat{S}_z | \psi_m \rangle \langle \psi_m | \hat{L}_z + 2\hat{S}_z | \psi_n \rangle}{W_n^0 - W_m^0}. \tag{3.33}$$

The comparison betwen Equations 3.32 and 3.33 gives:

$$W_n^{(1)} = \beta \langle \psi_n | \hat{L}_z + 2\hat{S}_z | \psi_n \rangle$$

$$W_n^{(2)} = \sum_m{}' \frac{\beta^2 \langle \psi_n | \hat{L}_z + 2\hat{S}_z | \psi_m \rangle \langle \psi_m | \hat{L}_z + 2\hat{S}_z | \psi_n \rangle}{W_n^0 - W_m^0}.$$

Equation 3.33 shows that the first-order Zeeman effect on the nth level involves the change in energy due to the magnetic field acting on that level alone, whilst *the second-order Zeeman effect arises from the change in energy due to mixing in, via the magnetic field, of levels which were not originally degenerate with the nth level.* Similar expressions to those given above are obtained when the level is degenerate in the absence of a magnetic field, but in this case the degeneracy is normally removed resulting in a number of different first- and second-order Zeeman coefficients, corresponding to the degeneracy of the original level. The ability to calculate the first- and second-order Zeeman coefficients from a given set of wave-functions obviously enables a comparison to be made between a proposed electronic structure of a molecule and the observed magnetic susceptibility. This treatment will be elaborated upon in Chapters 4 and 5.

The effect of spin–orbit coupling on crystal field terms will be discussed in later chapters and the terminology first- and second-order spin–orbit coupling will be used. These terms are analogous to the first- and second-order Zeeman effect in that the first-order term will be the effect of spin–orbit coupling within a given level (which may be degenerate), whilst the second-order term is that involving spin–orbit coupling between levels which were not initially degenerate.

3.4. The relationship between the g values from electron spin resonance and the wave-functions of a system

A simple relationship between the first- and second-order Zeeman coefficients and the wave-functions within an energy level was demonstrated in the last section. Similar expressions to these may also be derived which relate the g values, obtained from electron spin resonance experiments, to the wave-functions of the lowest energy levels. Since the expressions are most commonly used and easily derived for a one unpaired electron system, this case will be used to illustrate the method of calculation. Multi-electron systems may be treated in a similar manner.

Electron spin resonance spectra are usually interpreted by means of a spin Hamiltonian rather than the actual Hamiltonian for the system [3, 4]. The spin Hamiltonian is an artificial concept which involves only spin operators and allows the wave-functions for the system to be expressed in terms of spin functions only. Any orbital angular momentum which the state possesses is allowed for by making the g value different from 2, the value for a free electron. The spin Hamiltonian and spin functions appropriate to a one unpaired electron system are:

$$\mathcal{H}_s = g_z\beta H_z\hat{S}_z + g_x\beta H_x\hat{S}_x + g_y\beta H_y\hat{S}_y \tag{3.34}$$

and

$$|\tfrac{1}{2}\rangle, |-\tfrac{1}{2}\rangle,$$

where x, y, and z refer to the principal molecular directions. The energy matrix resulting from (3.34) is:

$$
\begin{array}{cc}
 & |\tfrac{1}{2}\rangle \qquad\qquad\qquad |-\tfrac{1}{2}\rangle \\
\begin{array}{c} \langle\tfrac{1}{2}| \\[2em] \langle-\tfrac{1}{2}| \end{array}
&
\left[
\begin{array}{cc}
\dfrac{g_z\beta H_z}{2} & \dfrac{g_x\beta H_x}{2} - \dfrac{ig_y\beta H_y}{2} \\[1.5em]
\dfrac{g_x\beta H_x}{2} + \dfrac{ig_y\beta H_y}{2} & \dfrac{-g_z\beta H_z}{2}
\end{array}
\right].
\end{array}
\tag{3.35}
$$

The methods whereby the principal g values are obtained from electron spin resonance spectra are beyond the scope of this book. Since we are only interested in showing the relationship between the g values and the wave-functions of the ground state on our system, it is sufficient to know that it can be done.

If we now transfer our attention to the effect of a magnetic field on the actual wave-functions of the ground state of the system we have the Hamiltonian:

$$\mathcal{H} = \beta H_z(\hat{L}_z + 2\hat{S}_z) + \beta H_x(\hat{L}_x + 2\hat{S}_x) + \beta H_y(\hat{L}_y + 2\hat{S}_y) \tag{3.36}$$

whilst the wave-functions before the application of the magnetic field may be expressed as $|\psi+\rangle$ and $|\psi-\rangle$, where $|\psi+\rangle$ is the wave-function which is associated mainly with the spin function $m_s = +\tfrac{1}{2}$ and similarly $|\psi-\rangle$ with $m_s = -\tfrac{1}{2}$. The energy matrix given by (3.36) is:

$$
\begin{array}{cc}
 & |\psi+\rangle \qquad\qquad\qquad\qquad |\psi-\rangle \\
\begin{array}{c} \langle\psi+| \\[3em] \langle\psi-| \end{array}
&
\left[
\begin{array}{cc}
\langle\psi+|\hat{L}_z + 2\hat{S}_z|\psi+\rangle\beta H_z & \begin{array}{l}\langle\psi+|\hat{L}_x + 2\hat{S}_x|\psi-\rangle\beta H_x \\ + \langle\psi+|\hat{L}_y + 2\hat{S}_y|\psi-\rangle\beta H_y\end{array} \\[2.5em]
\begin{array}{l}\langle\psi-|\hat{L}_x + 2\hat{S}_x|\psi+\rangle\beta H_x \\ + \langle\psi-|\hat{L}_y + 2\hat{S}_y|\psi+\rangle\beta H_y\end{array} & \langle\psi-|\hat{L}_z + 2\hat{S}_z|\psi-\rangle\beta H_z
\end{array}
\right].
\end{array}
\tag{3.37}
$$

If the energy matrices (3.35) and (3.37) are to give the same energy levels in the magnetic field, which they must if the spin Hamiltonian is valid, then like matrix elements may be equated. This then leads to the following relationships between the principal g values and wave-functions for the system:

$$\left.\begin{array}{c} g_z = 2\langle\psi +|\hat{L}_z + 2\hat{S}_z|\psi +\rangle \\ g_x = 2\langle\psi +|\hat{L}_x + 2\hat{S}_x|\psi -\rangle \\ g_y = 2i\langle\psi +|\hat{L}_y + 2\hat{S}_y|\psi -\rangle \end{array}\right\}. \tag{3.38}$$

Since the ground state wave-functions, $|\psi\pm\rangle$, usually involve the admixture of excited states the observed g values can be used to estimate the mixing coefficients using Equations 3.38. An example of this is shown for a copper(II) system in Chapter 6.

3.5. Simultaneous perturbations

In some systems there may be two or more factors which perturb the system, e.g. a crystal field and spin–orbit coupling. Under these circumstances it is sometimes permissible to treat these perturbations separately as consecutive perturbations, e.g. solve the problem for the perturbation by the crystal field and then perturb these solutions by spin–orbit coupling to give a final set of energy levels and wave-functions. However, we must exercise care when doing this since this procedure is only valid when the effects of the second perturbation are much smaller than those of the first. If this condition is not satisfied then the two perturbations must be applied simultaneously. The main problem facing us is how small is 'much smaller than'. In general, individual cases should be treated on their respective merits but the following hypothetical problem will suffice to illustrate the consequences of using the two procedures above. If in doubt it is safer to apply the perturbations simultaneously, although this may involve more labour.

Consider the case where there is a single electron in the degenerate pair of orbitals d_{xz}, d_{yz} in the absence of a crystal field and spin–orbit coupling. Further assume that the crystal field is such that it removes the degeneracy of the orbitals to give d_{yz} at an energy Δ above d_{xz}. Let us now calculate the energies of these orbitals when the two perturbations are applied (1) consecutively and (2) simultaneously.

(1) The wave-functions involved are $|d_{xz}, \frac{1}{2}\rangle$, $|d_{xz}, -\frac{1}{2}\rangle$, $|d_{yz}, \frac{1}{2}\rangle$ and $|d_{yz}, -\frac{1}{2}\rangle$. Applying the crystal field perturbation first we have the wave-functions $|d_{xz}, \frac{1}{2}\rangle$, $|d_{xz}, -\frac{1}{2}\rangle$ at energy zero, whilst $|d_{yz}, \frac{1}{2}\rangle$ and $|d_{yz}, -\frac{1}{2}\rangle$ are at energy Δ. For simplicity we will now calculate the energies of only the lowest level under the action of spin–orbit coupling. The problem is that of a degenerate state since there is two-fold spin degeneracy, i.e. $m_s = \pm\frac{1}{2}$. Thus to determine the first-order energies after perturbation by spin–orbit coupling we need to solve the secular determinant of the type (3.24). This requires evaluating the matrix elements $\langle d_{xz}, \pm\frac{1}{2}|\xi\hat{l}\cdot\hat{s}|d_{xz}, \pm\frac{1}{2}\rangle$ and $\langle d_{xz}, \pm\frac{1}{2}|\xi\hat{l}\cdot\hat{s}|d_{xz}, \mp\frac{1}{2}\rangle$. This may be done using

Equations 3.27 and 3.30 and the functions given in Table 4.1. All four matrix elements involved are zero which immediately tells us that spin–orbit coupling to the first order leaves these two functions unchanged, i.e. $E_i^{(1)}$ in Equation 3.26 is zero for each component. This now leaves the third term in 3.26, the second-order correction to the energy, to be determined. We will only calculate this for $|d_{xz}, \frac{1}{2}\rangle$ since an identical result is obtained for the other component. The only non-zero matrix elements in the present system are $\langle d_{xz}, \frac{1}{2}|\zeta \hat{l}.\hat{s}|d_{yz}, \frac{1}{2}\rangle = -i\zeta/2$ and $\langle d_{yz}, \frac{1}{2}|\zeta \hat{l}.\hat{s}|d_{xz}, \frac{1}{2}\rangle = i\zeta/2$. Substitution in Equation 3.26 gives the energy of what is primarily the $|d_{xz}, \frac{1}{2}\rangle$ function as:

$$E = \frac{-\zeta^2}{4\Delta}. \tag{3.39}$$

(2) Here we are dealing with a four-fold degenerate problem (i.e. two-fold spin and two-fold orbital degeneracy) with the perturbing operator:

$$\mathscr{H}^{(1)} = V + \zeta \hat{l}.\hat{s}$$

where V represents the crystal field potential. Evaluation of the matrix elements of the type

$$\langle \psi_n|\mathscr{H}^{(1)}|\psi_m\rangle = \langle \psi_n|V|\psi_m\rangle + \langle \psi_n|\zeta \hat{l}.\hat{s}|\psi_m\rangle$$

enables us to set up the secular determinant and solve for the energies:

	$\lvert d_{xz}, \frac{1}{2}\rangle$	$\lvert d_{yz}, \frac{1}{2}\rangle$	$\lvert d_{xz}, -\frac{1}{2}\rangle$	$\lvert d_{yz}, -\frac{1}{2}\rangle$		
$\langle d_{xz}, \frac{1}{2}\rvert$	$0 - E^{(1)}$	$\dfrac{-i\zeta}{2}$	0	0		
$\langle d_{yz}, \frac{1}{2}\rvert$	$\dfrac{i\zeta}{2}$	$\Delta - E^{(1)}$	0	0	$= 0$	(3.40)
$\langle d_{xz}, -\frac{1}{2}\rvert$	0	0	$0 - E^{(1)}$	$\dfrac{i\zeta}{2}$		
$\langle d_{yz}, -\frac{1}{2}\rvert$	0	0	$\dfrac{-i\zeta}{2}$	$\Delta - E^{(1)}$		

Solution of either of the sub-determinants gives:

$$E^{(1)} = \tfrac{1}{2}[\Delta \pm (\Delta^2 + \zeta^2)^{\frac{1}{2}}]. \tag{3.41}$$

The lower root, which corresponds to the state which is primarily $|d_{xz}, \frac{1}{2}\rangle$ (at least when $\Delta \geqslant \zeta$), gives:

$$E = \tfrac{1}{2}[\Delta - (\Delta^2 + \zeta^2)^{\frac{1}{2}}]. \tag{3.42}$$

This solution involving the simultaneous perturbation, is often referred to as an exact solution to the problem (*exact in the context of the perturbations being applied*), whereas that from the stepwise perturbation process is an approximate solution. If the stepwise perturbation process is a good approximation then Equations 3.42 and 3.39 should give the same results. Clearly this is not so for all values of ζ and Δ, but the energies do approach each other as ζ becomes progressively smaller compared with Δ.

Table 3.1.

	$E = \frac{1}{2}[\Delta - (\Delta^2 + \zeta^2)^{\frac{1}{2}}]$	$E = \dfrac{-\zeta^2}{4\Delta}$
$\Delta = \zeta/2$	$-0 \cdot 618\Delta$	$-1 \cdot 000\Delta$
$\Delta = \zeta$	$-0 \cdot 207\Delta$	$-0 \cdot 250\Delta$
$\Delta = 2\zeta$	$-0 \cdot 090\Delta$	$-0 \cdot 063\Delta$
$\Delta = 4\zeta$	$-0 \cdot 0156\Delta$	$-0 \cdot 0156\Delta$

Table 3.1 gives a comparison of the energies of the lowest level calculated by both methods. As expected when the effect of the crystal field is half that of the spin–orbit coupling the agreement is extremely poor. Increasing Δ relative to ζ gradually improves the agreement and in this particular example when Δ is four times the spin–orbit coupling constant the two methods give the same result.

REFERENCES

1. H. Eyring, J. Walter and G.E. Kimball (1964). *Quantum Chemistry*, Wiley, New York. Chapter VII.
2. L. Pauling and E.B. Wilson, Jun. (1935). *Introduction to Quantum Mechanics*, McGraw-Hill, New York. Chapter VI.
3. M.H.L. Pryce (1950). *Proc. Phys. Soc. (London),* **A63**, 25; A. Abragam and M.H.L. Pryce (1951). *Proc. Roy. Soc. (London),* **A205**, 135.
4. A. Carrington and A.D. McLachlan (1967). *Introduction to Magnetic Resonance,* Harper International Edition, New York. 133–136; B.R. McGarvey (1966). *Transition Metal Chemistry,* Edited by R.L. Carlin, Edward Arnold Ltd., London. **3**, 111.

4

The magnetic properties of transition metal ions in cubic crystal fields

4.1. Introduction

In Chapter 2 it was shown that, when a transition metal ion is subjected to an octahedral or tetrahedral crystal field, some of the five-fold degeneracy of the d-orbitals is removed. This results in new energy levels for the system, i.e. the crystal field terms. For example in a weak octahedral crystal field this leads to the free-ion ^2D term being split into the $^2T_2$ and ^2E crystal field terms.* These cubic field terms provide the starting point for the calculation and discussion of the magnetic susceptibilities of transition metal ions in cubic crystal fields to be undertaken in this chapter. Thus a great deal of space will be devoted to a discussion of the effects, on the crystal field terms, of consecutive perturbation by spin–orbit coupling and magnetic fields.

4.2. The quenching of the orbital contribution to the magnetic susceptibility

If the electrons in a transition metal ion could be considered to be free, i.e. there is no influence from crystal fields or spin–orbit coupling, then the effective magnetic moment would simply be the sum of the contributions from the spin and orbital angular momenta: .

$$\mu = [4S(S + 1) + L(L + 1)]^{\frac{1}{2}} \text{ B.M.} \qquad (4.1)$$

Reference to Table 1.2 shows that nearly all the observed magnetic moments of the first transition series ions deviate from this value, usually being much closer to the spin-only value. This effect is commonly known as 'quenching' of the

* Strictly the terms should be $^2T_{2g}$ and ^2E_g in an octahedral field but for convenience throughout this chapter the g subscript will be omitted.

orbital contribution, and in transition metal complexes it can be attributed to the effects of the crystal field and spin–orbit coupling.

The 'quenching' of the orbital contribution by the crystal field alone can be illustrated both qualitatively and quantitatively from a consideration of a set of d-orbitals. In the qualitative approach, previously mentioned in Chapter 1, it is assumed that for an electron in a particular orbital to have orbital angular momentum about a given axis, it must be possible, by rotation about that axis, to transform the orbital into an *equivalent* and *degenerate* orbital which does not already contain an electron with the same spin. Such a transformation can be considered to allow the electron to rotate about the axis, i.e. the electron has an angular momentum about the axis. In a free ion a rotation of 45° about the z-axis will convert the d_{xy} orbital into the $d_{x^2-y^2}$ orbital, whilst a rotation of 90° transforms the d_{xz} into the d_{yz} orbital. Hence we would expect an orbital contribution to the magnetic moment from an electron in either of these pairs of orbitals. The d_{z^2} orbital, however, cannot be transformed into any of the other orbitals by rotation about the z-axis and thus an electron in this orbital cannot give an orbital contribution to the magnetic moment. When the ion is placed in a cubic crystal field the degeneracy of the d-orbitals is removed giving the t_2 set (d_{xy}, d_{yz}, d_{xz}) and the e set $(d_{z^2}, d_{x^2-y^2})$, separated by $10Dq$. The consequence of this is that the degeneracy of the d_{xy} and $d_{x^2-y^2}$ orbitals is destroyed and hence the orbital contribution from an electron in either one of these orbitals is lost. However, the d_{xz} and d_{yz} orbitals remain degenerate and consequently are capable of giving an orbital contribution to the magnetic moment. It is worthwhile noting that, since the d_{z^2} and $d_{x^2-y^2}$ orbitals cannot be transformed into each other by rotation about any axis, there can be no orbital angular momentum associated with this e set of orbitals. Table 1.3 summarizes the d-electronic configurations which will or will not have their orbital contributions quenched.

In the quantitative approach the effect of a magnetic field on the orbital degeneracies of a set of d-orbitals will be calculated. This calculation will provide a useful and relatively simple example of the application of perturbation theory. The orbital parts of the wave-functions for the d-orbitals, expressed in their real forms are:

$$d_{z^2} = |0\rangle$$
$$d_{x^2-y^2} = \sqrt{\tfrac{1}{2}}(|2\rangle + |-2\rangle)$$
$$d_{xy} = i\sqrt{\tfrac{1}{2}}(|-2\rangle - |2\rangle)$$
$$d_{xz} = \sqrt{\tfrac{1}{2}}(|-1\rangle - |1\rangle)$$
$$d_{yz} = i\sqrt{\tfrac{1}{2}}(|-1\rangle + |1\rangle).$$

The particular problem we wish to solve is that of a degenerate state perturbed by a magnetic field, i.e. we need to solve an equation of the type

$$(\mathcal{H}_0 + \mathcal{H}^{(1)})\psi = E\psi, \tag{4.2}$$

where $\mathcal{H}_0$ is the Hamiltonian corresponding to the hydrogen-like atom which resulted in our set of degenerate d-orbitals.

$\mathcal{H}^{(1)}$ is the Hamiltonian expressing the perturbation by the magnetic field, which in our specific case we have chosen to simplify to $\hat{l}_z\beta H$ (cf. Chapter 3 for the magnetic moment operator), since for the present we are not concerned with the spin contribution.

E and ψ are the energies and wave-functions which will satisfy the above equation.

The first-order approximations to the energies E can be found using perturbation theory. This treatment requires the solution of the secular determinant in which the matrix elements I_{nm} are $\langle d_n|\hat{l}_z\beta H|d_m\rangle$. The evaluation of one of these matrix elements is given in detail; the evaluation of the remainder forms a useful exercise.

Example $\langle d_{x^2-y^2}|\hat{l}_z\beta H|d_{xy}\rangle$

The first step is to operate with $\hat{l}_z\beta H$ on the wave-function $|d_{xy}\rangle$, then premultiply by $\langle d_{x^2-y^2}|$ and evaluate the resulting integral:

$$\hat{l}_z\beta H|d_{xy}\rangle = \hat{l}_z\beta Hi\sqrt{\tfrac{1}{2}}(|-2\rangle - |2\rangle)$$

$$= \beta Hi\sqrt{\tfrac{1}{2}}(\hat{l}_z|-2\rangle - \hat{l}_z|2\rangle)$$

$$= \beta Hi\sqrt{\tfrac{1}{2}}(-2|-2\rangle - 2|2\rangle)$$

$$= -\beta H2i\sqrt{\tfrac{1}{2}}(|-2\rangle + |2\rangle)$$

$$= -\beta H2i|d_{x^2-y^2}\rangle$$

$$\langle d_{x^2-y^2}|\hat{l}_z\beta H|d_{xy}\rangle = \langle d_{x^2-y^2}|(-\beta H2i\, d_{x^2-y^2})\rangle$$

$$= -\beta H2i\langle d_{x^2-y^2}|d_{x^2-y^2}\rangle$$

$$= -\beta H2i, \text{ since } \langle d_{x^2-y^2}|d_{x^2-y^2}\rangle = 1.$$

Table 4.1. *Wave-functions obtained from real d-atomic orbitals using $\hat{l}_z, \hat{l}_x, \hat{l}_y$.*

Orbital	Result of operating with		
	$\hat{l}_z$	$\hat{l}_x$	$\hat{l}_y$
d_{z^2}	$0\, d_{z^2}$	$-i\sqrt{3}\, d_{yz}$	$i\sqrt{3}\, d_{xz}$
$d_{x^2-y^2}$	$2i\, d_{xy}$	$-i\, d_{yz}$	$-i\, d_{xz}$
d_{xy}	$-2i\, d_{x^2-y^2}$	$i\, d_{xz}$	$-i\, d_{yz}$
d_{xz}	$i\, d_{yz}$	$-i\, d_{xy}$	$-i\sqrt{3}\, d_{z^2} + i\, d_{x^2-y^2}$
d_{yz}	$-i\, d_{xz}$	$i\, d_{x^2-y^2} + i\sqrt{3}\, d_{z^2}$	$i\, d_{xy}$

The results of operating with $\hat{l}_z$, $\hat{l}_x$ or $\hat{l}_y$ on the real form of the d atomic orbitals are given in Table 4.1. Using this table we have the following determinant:

	$|d_{xz}\rangle$	$|d_{yz}\rangle$	$|d_{xy}\rangle$	$|d_{x^2-y^2}\rangle$	$|d_{z^2}\rangle$
$\langle d_{xz}|$	$0 - E$	$-i\beta H$	0	0	0
$\langle d_{yz}|$	$i\beta H$	$0 - E$	0	0	0
$\langle d_{xy}|$	0	0	$0 - E$	$2i\beta H$	0
$\langle d_{x^2-y^2}|$	0	0	$-2i\beta H$	$0 - E$	0
$\langle d_{z^2}|$	0	0	0	0	$0 - E$

$$= 0.$$

$$(4.3)$$

Solving each sub-determinant we find that the energies which would arise from the interaction of the orbital angular momentum with the magnetic field are:

$$E = \pm 2\beta H, \ \pm 1\beta H \text{ and } 0$$

(see Fig. 4.1). Each of the energies we have calculated correspond to the first-order Zeeman terms in the expansion

$$W_i = W_i^0 + W_i^{(1)}H + W_i^{(2)}H^2 + \ldots$$

and hence we can readily calculate the orbital contribution to the magnetic susceptibility using Van Vleck's equation, i.e.

$$\chi = \frac{N((2\beta)^2 + \beta^2 + 0^2 + (-\beta)^2 + (-2\beta)^2)}{5kT} = \frac{2N\beta^2}{kT}.$$

$$(4.4)$$

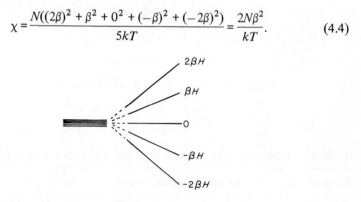

Magnetic field H

Figure 4.1. The variation of the energies of the d-orbitals in a magnetic field when only the orbital angular momentum is considered.

If we now consider the situation where an octahedral crystal field has split the d-orbitals into a lower t_2 set and an upper e set, the orbital contribution to the magnetic susceptibility comes almost entirely from the t_2 set, since these are the only thermally populated orbitals at normal temperatures. The energy levels which result from our perturbation can be obtained to the first order from the following secular determinant:

	$\lvert d_{xz} \rangle$	$\lvert d_{yz} \rangle$	$\lvert d_{xy} \rangle$	
$\langle d_{xz} \rvert$	$0 - E$	$-i\beta H$	0	
$\langle d_{yz} \rvert$	$i\beta H$	$0 - E$	0	$= 0,$ (4.5)
$\langle d_{xy} \rvert$	0	0	$0 - E$	

whence $E = \pm \beta H$ and 0.

However, there is a non-zero matrix element between $\lvert d_{xy} \rangle$ from the t_2 set and $\lvert d_{x^2-y^2} \rangle$ from the e set. This gives a second-order correction to the energy of

$$\frac{\langle d_{xy} \lvert \hat{l}_z \beta H \rvert d_{x^2-y^2} \rangle \langle d_{x^2-y^2} \lvert \hat{l}_z \beta H \rvert d_{xy} \rangle}{-10Dq} = \frac{-4\beta^2 H^2}{10Dq}.$$

Hence the susceptibility is

$$\chi = \frac{N \left(\dfrac{2\beta^2}{kT} + \dfrac{8\beta^2}{10Dq} \right)}{3} = \frac{2N\beta^2}{3kT} + \frac{8N\beta^2}{30Dq}. \qquad (4.6)$$

The T.I.P. term will be much smaller than the first-order term because of the magnitude of $10Dq$, thus the calculation demonstrates quantitatively the quenching of the orbital contribution to the magnetic susceptibility by comparison with Equation 4.4.

4.3. The magnetic properties of transition metal ions in weak octahedral and tetrahedral crystal fields

In Chapter 2 the weak crystal field terms, and the corresponding wave-functions, which arise from various d-electron configurations have been given. The quenching of the orbital contribution to the magnetic moment due to the action of the crystal field was discussed in the previous section. However, spin–orbit coupling is another perturbation which can affect the orbital contribution and we will discuss, and in a few cases calculate, its effect on the magnetic susceptibilities of transition metal complexes. Anticipating the final results we find that the magnetic susceptibilities of ions with T ground terms are very dependent on the effects of spin–orbit coupling because in the first-order approximation this perturbation removes some of the degeneracy of the term, giving levels which may be separated by energies of the order of kT. The behaviour of magnetic susceptibilities under these circumstances was discussed in Chapter 1. Since there is no orbital angular momentum associated with the A

and E terms which arise for d-transition metal ions, spin–orbit coupling cannot raise any of the degeneracy and thus to a first-order approximation it has no effect on the magnetic properties. In fact, because there is no orbital angular momentum associated with these terms, we should observe the spin-only magnetic moment to this order of approximation. Some orbital contribution can, however, be introduced in a second-order treatment due to 'mixing in' of excited orbitally degenerate terms.

The details of the calculation of the magnetic susceptibilities of transition metal complexes in cubic crystal fields will be illustrated by applying perturbation theory to the $^2T_2$ and $^3A_2$ ground terms. The results for the other ground terms which arise are calculated in the same way, but only final results will be quoted and discussed here.

4.3.1. THE MAGNETIC SUSCEPTIBILITY OF THE $^2T_2$ TERM ARISING FROM A FREE ION ^2D TERM.

The magnetic susceptibility of this term can conveniently be calculated by considering the effects of spin–orbit coupling and a magnetic field as sequential perturbations. We are able to do this because the effects of spin–orbit coupling give rise to splittings of the $^2T_2$ term of at least 100 cm^{-1}, whereas the effects of the magnetic field under most laboratory conditions will only give energy splittings of the order of 1 cm^{-1}

The wave-functions contained within the $^2T_2$ term, which arises from a free ion ^2D term, are all associated with the quantum numbers $L = 2$ and $S = \frac{1}{2}$. Because the effects of the spin–orbit coupling and magnetic moment operators can be evaluated in terms of quantum numbers, L, S, M_L, and M_S we can specify our wave-functions in terms of these quantum numbers, i.e. $|L, M_L, S, M_S\rangle$. However, in the following calculation we will only be concerned with evaluating matrix elements between wave-functions within the $^2T_2$ term for which L and S are constant and hence we will abbreviate our wave-function to $|M_L, M_S\rangle$. A typical matrix element will be $\langle M'_L, M'_S|\lambda\hat{L} . \hat{S}|M_L, M_S\rangle$.

The full set of wave-functions for the $^2T_2$ term are:

$$\psi_1 = |1, \tfrac{1}{2}\rangle.$$
$$\psi_2 = |1, -\tfrac{1}{2}\rangle.$$
$$\psi_3 = |-1, \tfrac{1}{2}\rangle.$$
$$\psi_4 = |-1, -\tfrac{1}{2}\rangle.$$
$$\psi_5 = \sqrt{\tfrac{1}{2}}(|2, \tfrac{1}{2}\rangle - |-2, \tfrac{1}{2}\rangle).$$
$$\psi_6 = \sqrt{\tfrac{1}{2}}(|2, -\tfrac{1}{2}\rangle - |-2. -\tfrac{1}{2}\rangle).$$

The first stage in the calculation is to find the energy levels, and the corresponding wave-functions, which arise from the action of spin–orbit coupling. Since we are dealing with a degenerate state we need to construct the secular determinant involving the six wave-functions. The solution of this determinant will give the first-order correction to the energy of the $^2T_2$ term and also the zeroth order wave-functions. In order to construct this determinant, matrix elements of the type $\langle\psi_i|\lambda\hat{L}\cdot\hat{S}|\psi_j\rangle$ have to be evaluated. Two such matrix elements will be evaluated in detail, the evaluation of the remainder form a useful exercise for the reader. A reminder of the expressions for the spin–orbit coupling operator in terms of the components of $\hat{L}$ and $\hat{S}$, and of the effects of operating with $\hat{L}$ and $\hat{S}$ on a wave-function, is appropriate at this point.

$$\lambda\hat{L}\cdot\hat{S} = \lambda(\hat{L}_z\hat{S}_z + \hat{L}_x\hat{S}_x + \hat{L}_y\hat{S}_y)$$

$$= \lambda\hat{L}_z\hat{S}_z + \frac{\lambda}{2}(\hat{L}_+\hat{S}_- + \hat{L}_-\hat{S}_+). \tag{4.7}$$

$$\left.\begin{aligned}
\hat{L}_z|L, M_L, S, M_S\rangle &= M_L|L, M_L, S, M_S\rangle \\
\hat{S}_z|L, M_L, S, M_S\rangle &= M_S|L, M_L, S, M_S\rangle \\
\hat{L}_+|L, M_L, S, M_S\rangle &= [L(L+1) - M_L(M_L+1)]^{\frac{1}{2}}|L, (M_L+1), S, M_S\rangle \\
\hat{L}_-|L, M_L, S, M_S\rangle &= [L(L+1) - M_L(M_L-1)]^{\frac{1}{2}}|L, (M_L-1), S, M_S\rangle \\
\hat{S}_+|L, M_L, S, M_S\rangle &= [S(S+1) - M_S(M_S+1)]^{\frac{1}{2}}|L, M_L, S, (M_S+1)\rangle \\
\hat{S}_-|L, M_L, S, M_S\rangle &= [S(S+1) - M_S(M_S-1)]^{\frac{1}{2}}|L, M_L, S, (M_S-1)\rangle
\end{aligned}\right\} \cdot \tag{4.8}$$

The value of L used in the above equations is that of the free ion term from which the wave-functions originate.

Example 1. $\langle\psi_2|\lambda\hat{L}\cdot\hat{S}|\psi_2\rangle$

Step 1. Write in full expressions for the wave-functions and expand the spin–orbit coupling operator.

$$\langle 1, -\tfrac{1}{2}|\lambda\hat{L}_z\hat{S}_z + \frac{\lambda}{2}(\hat{L}_+\hat{S}_- + \hat{L}_-\hat{S}_+)|1, -\tfrac{1}{2}\rangle,$$

Step 2. Separate the matrix element into its parts

$$\langle 1, -\tfrac{1}{2}|\lambda\hat{L}_z\hat{S}_z|1, -\tfrac{1}{2}\rangle + \left\langle 1, -\tfrac{1}{2}\left|\frac{\lambda}{2}\hat{L}_+\hat{S}_-\right|1, -\tfrac{1}{2}\right\rangle + \left\langle 1, -\tfrac{1}{2}\left|\frac{\lambda}{2}\hat{L}_-\hat{S}_+\right|1, -\tfrac{1}{2}\right\rangle.$$

Step 3. Evaluate each individual matrix element using the Equations 4.8. If we consider the matrix element $\langle 1, -\tfrac{1}{2}|\lambda\hat{L}_z\hat{S}_z|1, -\tfrac{1}{2}\rangle$, the first stage in the evaluation is to operate with $\hat{L}_z\hat{S}_z$ on the wave-function written on its right-hand side, i.e. $|1, -\tfrac{1}{2}\rangle$. The result of this is

$$1, -\tfrac{1}{2}|\lambda\hat{L}_z\hat{S}_z|1, -\tfrac{1}{2}\rangle = \langle 1, -\tfrac{1}{2}|\lambda(1)(-\tfrac{1}{2})|1, -\tfrac{1}{2}\rangle$$

$$= -\frac{\lambda}{2}\langle 1, -\tfrac{1}{2}|1, -\tfrac{1}{2}\rangle$$

$$= -\frac{\lambda}{2}, \text{ since } \langle 1, -\tfrac{1}{2}|1, -\tfrac{1}{2}\rangle = 1$$

$$-\tfrac{1}{2}\left|\frac{\lambda}{2}\hat{L}_-\hat{S}_+\right|1, -\tfrac{1}{2}\rangle = \left\langle 1, -\tfrac{1}{2}\left|\frac{\lambda}{2}[2(2+1) - 1(1-1)]^{\frac{1}{2}}[\tfrac{1}{2}(\tfrac{3}{2}) - (-\tfrac{1}{2})(-\tfrac{1}{2}+1)]^{\frac{1}{2}}\right|0, \tfrac{1}{2}\right.$$

$$= \frac{\lambda}{2} \times \sqrt{6} \times 1\langle 1, -\tfrac{1}{2}|0, \tfrac{1}{2}\rangle$$

$$= 0 \text{ since } \langle 1, -\tfrac{1}{2}|0, \tfrac{1}{2}\rangle = 0.$$

Similarly $\left\langle 1, -\tfrac{1}{2}\left|\frac{\lambda}{2}\hat{L}_+\hat{S}_-\right|1, -\tfrac{1}{2}\right\rangle = 0.$

$$\therefore \langle\psi_2|\lambda\hat{L}.\hat{S}|\psi_2\rangle = -\frac{\lambda}{2}.$$

Example 2. $\langle\psi_1|\lambda\hat{L}.\hat{S}|\psi_6\rangle.$

Following the individual steps given in Example 1:

$$\langle\psi_1|\lambda\hat{L}.\hat{S}|\psi_6\rangle = \left\langle 1, \tfrac{1}{2}\left|\lambda\hat{L}_z\hat{S}_z + \frac{\lambda}{2}(\hat{L}_+\hat{S}_- + \hat{L}_-\hat{S}_+)\right|\sqrt{\tfrac{1}{2}}(|2, -\tfrac{1}{2}\rangle - |-2, -\tfrac{1}{2}\rangle)\right\rangle$$

$$= \sqrt{\tfrac{1}{2}}\langle 1, \tfrac{1}{2}|\lambda\hat{L}_z\hat{S}_z|2, -\tfrac{1}{2}\rangle - \sqrt{\tfrac{1}{2}}\langle 1, \tfrac{1}{2}|\lambda\hat{L}_z\hat{S}_z|-2, -\tfrac{1}{2}\rangle$$

$$+ \sqrt{\tfrac{1}{2}}\left\langle 1, \tfrac{1}{2}\left|\frac{\lambda}{2}\hat{L}_+\hat{S}_-\right|2, -\tfrac{1}{2}\right\rangle - \sqrt{\tfrac{1}{2}}\left\langle 1, \tfrac{1}{2}\left|\frac{\lambda}{2}\hat{L}_+\hat{S}_-\right|-2, -\tfrac{1}{2}\right\rangle$$

$$+ \sqrt{\tfrac{1}{2}}\left\langle 1, \tfrac{1}{2}\left|\frac{\lambda}{2}\hat{L}_-\hat{S}_+\right|2, -\tfrac{1}{2}\right\rangle - \sqrt{\tfrac{1}{2}}\left\langle 1, \tfrac{1}{2}\left|\frac{\lambda}{2}\hat{L}_-\hat{S}_+\right|-2, -\tfrac{1}{2}\right\rangle.$$

Of these six matrix elements only the non-zero one will be evaluated in detail:

$$\sqrt{\tfrac{1}{2}}\left\langle 1, \tfrac{1}{2}\left|\frac{\lambda}{2}\hat{L}_-\hat{S}_+\right|2, -\tfrac{1}{2}\right\rangle$$

$$= \sqrt{\tfrac{1}{2}}\left\langle 1, \tfrac{1}{2}\left|\frac{\lambda}{2}[2(2+1) - 2(2-1)]^{\frac{1}{2}}[\tfrac{1}{2}(\tfrac{3}{2}) - (-\tfrac{1}{2})(-\tfrac{1}{2}+1)]^{\frac{1}{2}}\right|1, \tfrac{1}{2}\right\rangle$$

$$= \frac{\lambda}{\sqrt{2}}\langle 1, \tfrac{1}{2}|1, \tfrac{1}{2}\rangle = \frac{\lambda}{\sqrt{2}},$$

$$\therefore \langle\psi_1|\lambda\hat{L}.\hat{S}|\psi_6\rangle = \frac{\lambda}{\sqrt{2}}.$$

Evaluation of all the possible matrix elements for the $^2T_2$ term gives the following secular determinant:

$$
\begin{array}{c|cccccc}
 & |\psi_1\rangle & |\psi_2\rangle & |\psi_3\rangle & |\psi_4\rangle & |\psi_5\rangle & |\psi_6\rangle \\
\hline
\langle\psi_1| & \dfrac{\lambda}{2}-E & 0 & 0 & 0 & 0 & \dfrac{\lambda}{\sqrt{2}} \\
\langle\psi_2| & 0 & -\dfrac{\lambda}{2}-E & 0 & 0 & 0 & 0 \\
\langle\psi_3| & 0 & 0 & -\dfrac{\lambda}{2}-E & 0 & 0 & 0 \\
\langle\psi_4| & 0 & 0 & 0 & \dfrac{\lambda}{2}-E & -\dfrac{\lambda}{\sqrt{2}} & 0 \\
\langle\psi_5| & 0 & 0 & 0 & -\dfrac{\lambda}{\sqrt{2}} & 0-E & 0 \\
\langle\psi_6| & \dfrac{\lambda}{\sqrt{2}} & 0 & 0 & 0 & 0 & 0-E \\
\end{array}
= 0. \quad (4.9)
$$

Using the rules for interchanging rows and columns in a determinant, (4.9) can be rearranged so that it breaks down into a number of sub-determinants which can easily be solved.

$$
\begin{array}{c|cc|cc|cc}
 & |\psi_2\rangle & |\psi_3\rangle & |\psi_4\rangle & |\psi_5\rangle & |\psi_1\rangle & |\psi_6\rangle \\
\hline
\langle\psi_2| & -\dfrac{\lambda}{2}-E & 0 & 0 & 0 & 0 & 0 \\
\langle\psi_3| & 0 & -\dfrac{\lambda}{2}-E & 0 & 0 & 0 & 0 \\
\hline
\langle\psi_4| & 0 & 0 & \dfrac{\lambda}{2}-E & -\dfrac{\lambda}{\sqrt{2}} & 0 & 0 \\
\langle\psi_5| & 0 & 0 & -\dfrac{\lambda}{\sqrt{2}} & 0-E & 0 & 0 \\
\hline
\langle\psi_1| & 0 & 0 & 0 & 0 & \dfrac{\lambda}{2}-E & \dfrac{\lambda}{\sqrt{2}} \\
\langle\psi_6| & 0 & 0 & 0 & 0 & \dfrac{\lambda}{\sqrt{2}} & 0-E \\
\end{array}
= 0. \quad (4.10)
$$

Each sub-determinant (which is surrounded by the solid black lines) is a factor of the whole determinant and hence is equal to zero. Thus

$$
\langle\psi_2| \quad \begin{array}{|c|} \hline -\dfrac{\lambda}{2}-E \\ \hline \end{array} = 0. \qquad (4.11(a))
$$

$$
\langle\psi_3| \quad \begin{array}{|c|} \hline -\dfrac{\lambda}{2}-E \\ \hline \end{array} = 0. \qquad (4.11(b))
$$

$$
\begin{array}{c}
\quad\quad |\psi_4\rangle \quad\quad |\psi_5\rangle \\
\begin{array}{cc}
\langle\psi_4| \\
\langle\psi_5|
\end{array}
\begin{vmatrix}
\dfrac{\lambda}{2}-E & -\dfrac{\lambda}{\sqrt{2}} \\[2mm]
-\dfrac{\lambda}{\sqrt{2}} & 0-E
\end{vmatrix} = 0.
\end{array} \qquad (4.11(c))
$$

$$
\begin{array}{c}
\quad\quad |\psi_1\rangle \quad\quad |\psi_6\rangle \\
\begin{array}{cc}
\langle\psi_1| \\
\langle\psi_6|
\end{array}
\begin{vmatrix}
\dfrac{\lambda}{2}-E & \dfrac{\lambda}{\sqrt{2}} \\[2mm]
\dfrac{\lambda}{\sqrt{2}} & 0-E
\end{vmatrix} = 0.
\end{array} \qquad (4.11(d))
$$

The solutions to the Equations 4.11 are

$$
\left.
\begin{aligned}
& E_1 = -\frac{\lambda}{2}, \\
& E_2 = -\frac{\lambda}{2}, \\
& \left(\frac{\lambda}{2}-E\right)(0-E)-\left(-\frac{\lambda}{\sqrt{2}}\right)\left(-\frac{\lambda}{\sqrt{2}}\right) = 0. \\
& \text{i.e.} \\
& E_3 = \lambda \quad \text{and} \quad E_4 = -\frac{\lambda}{2}, \\
& \left(\frac{\lambda}{2}-E\right)(0-E)-\left(\frac{\lambda}{\sqrt{2}}\right)\left(\frac{\lambda}{\sqrt{2}}\right) = 0. \\
& \text{i.e.} \\
& E_5 = \lambda \quad \text{and} \quad E_6 = -\frac{\lambda}{2}.
\end{aligned}
\right\} \qquad (4.12)
$$

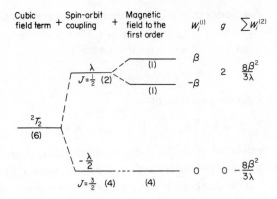

Figure 4.2. The splitting diagram for the $^2T_2$ term under the action of spin–orbit coupling and a magnetic field. (The figures in brackets under each energy level are the total degeneracies of the levels.)

The results of the calculation so far show that spin–orbit coupling splits the $^2T_2$ ground term into a four-fold degenerate level at $-(\lambda/2)$ and a doubly degenerate level at λ, relative to the energy of the $^2T_2$ term, see Fig. 4.2. However, the calculations of the effects of spin–orbit coupling are not yet complete, since we have to find the wave-functions associated with the above energy levels. The zeroth order wave-functions are obtained by substituting each of the energies into the set of simultaneous equations which gave rise to the secular determinant (see Chapter 3). The solution of these simultaneous equations for the ratios of the c_k's, coupled with the normalizing condition is sufficient to determine all the zeroth order wave-functions. This process is now illustrated with one example:

Consider the energy $E = \lambda$, then the set of simultaneous equations which results are:

$$
\left.
\begin{aligned}
-\frac{3\lambda}{2} c_2 &= 0 \\
-\frac{3\lambda}{2} c_3 &= 0 \\
-\frac{\lambda}{2} c_4 - \frac{\lambda}{\sqrt{2}} c_5 &= 0 \\
-\frac{\lambda}{\sqrt{2}} c_4 - \lambda c_5 &= 0 \\
-\frac{\lambda}{2} c_1 + \frac{\lambda}{\sqrt{2}} c_6 &= 0 \\
\frac{\lambda}{\sqrt{2}} c_1 - \lambda c_6 &= 0
\end{aligned}
\right\} .
\qquad (4.13)
$$

Where c_k is the coefficient of ψ_k in the zeroth order wave-function. This set of equations is satisfied by $c_2 = c_3 = 0$ and $c_6/c_1 = \sqrt{\frac{1}{2}}$, $c_4/c_1 = 1$, $c_5/c_1 = -\sqrt{\frac{1}{2}}$. Thus the zeroth order wave-function for this level is

$$\Psi = N[\psi_1 + \sqrt{\tfrac{1}{2}}\psi_6 + \psi_4 - \sqrt{\tfrac{1}{2}}\psi_5]$$

where N is the normalizing coefficient.

$$\therefore \ \Psi = \sqrt{\tfrac{1}{3}}[\psi_1 + \sqrt{\tfrac{1}{2}}\psi_6 + \psi_4 - \sqrt{\tfrac{1}{2}}\psi_5].$$

A more convenient set of zeroth order wave-functions, at least as far as the perturbation by a magnetic field is concerned, can be obtained by proceeding as follows. Instead of working with the complete determinant (4.10), we can obtain zeroth order wave-functions by substituting the energies we obtained from each sub-determinant (4.11) into the particular sub-determinant from which they were obtained.

Example 1:

Using (4.11(a)),

$$\begin{array}{c|c} & |\psi_2\rangle \\ \hline \langle\psi_2| & -\dfrac{\lambda}{2} - E \end{array} = 0,$$

for which $E_1 = -\lambda/2$. Substituting this value of E_1 into the equation involving c_k which would give this determinant we have

$$\left[-\frac{\lambda}{2} - \left(-\frac{\lambda}{2}\right)\right]c_2 = 0$$

i.e.

$$0c_2 = 0$$

From this result c_2 is indeterminate but it does tell us that an appropriate zeroth order wave-function for this particular energy involves only ψ_2, and since we are going to normalize to unity we can put $c_2 = 1$. Hence our zeroth order wave-function is $\phi_2 = \psi_2$.

Example 2.

Consider (4.11(d)),

$$\begin{array}{c|c c} & |\psi_1\rangle & |\psi_6\rangle \\ \hline \langle\psi_1| & \dfrac{\lambda}{2} - E & \dfrac{\lambda}{\sqrt{2}} \\ \langle\psi_6| & \dfrac{\lambda}{\sqrt{2}} & 0 - E \end{array} = 0,$$

from which we obtained $E_5 = \lambda$ and $E_6 = -(\lambda/2)$. If E_5 is substituted into the simultaneous equations which would give this determinant we have

$$\left[\frac{\lambda}{2} - \lambda\right] c_1 + \frac{\lambda}{\sqrt{2}} c_6 = 0$$

i.e.

$$\frac{\lambda}{\sqrt{2}} c_1 - \lambda c_6 = 0.$$

These equations give

$$\frac{c_6}{c_1} = \frac{1}{\sqrt{2}}$$

Thus a zeroth order wave-function corresponding to energy E_5 is

$$\Psi_1 = N[\psi_1 + \sqrt{\tfrac{1}{2}}\psi_6]$$

or

$$\Psi_1 = \sqrt{\tfrac{2}{3}}[\psi_1 + \sqrt{\tfrac{1}{2}}\psi_6].$$

Employing this method for all the other sub-determinants we find that the wave-functions are grouped in the following way.

At energy $-(\lambda/2)$ with respect to the $^2T_2$ level

$$\left.\begin{aligned}
\phi_1 &= \sqrt{\tfrac{1}{3}}[\psi_1 - \sqrt{2}\psi_6] \\
\phi_2 &= \psi_2 \\
\phi_3 &= \psi_3 \\
\phi_4 &= \sqrt{\tfrac{1}{3}}[\psi_4 + \sqrt{2}\psi_5]
\end{aligned}\right\} \tag{4.14}$$

and at energy λ with respect to the $^2T_2$ level

$$\left.\begin{aligned}
\Psi_1 &= \sqrt{\tfrac{2}{3}}[\psi_1 + \sqrt{\tfrac{1}{2}}\psi_6] \\
\Psi_2 &= \sqrt{\tfrac{2}{3}}[\psi_4 - \sqrt{\tfrac{1}{2}}\psi_5]
\end{aligned}\right\} \tag{4.15}$$

A more complete description of the wave-functions describing these two energy levels would involve 'mixing-in' from the higher ^2E level giving correction terms of the form

$$\frac{\langle\psi(^2E)|\lambda\hat{L}.\hat{S}|\psi(^2T_2)\rangle}{E(^2T_2) - E(^2E)} . \psi(^2E).$$

However, since these terms are inversely proportional to $10Dq$ they are normally neglected in comparison with the first-order effect when calculating magnetic susceptibilities. *In fact it must be emphasized at this point that all the magnetic susceptibilities calculated for T ground terms in this chapter ignore any contributions from excited terms.* We will see later that when there is no first-order effect on the ground state due to spin–orbit coupling, then the second-order effects are indeed very important. The wave-functions (4.14) and (4.15) and their corresponding energies now provide a starting point for calculating the magnetic susceptibility by treating the magnetic field as a perturbation on the problem we have just solved.

The term in the Hamiltonian expressing the perturbation due to the magnetic field is $\hat{\mu}H$, where $\hat{\mu}$ is the magnetic moment operator, $(\hat{L} + 2\hat{S})\beta$. Since the systems we are dealing with in this chapter have cubic symmetry they are isotropic and the average susceptibility is equal to that for any one of the principal susceptibilities of the molecule. Thus we need only calculate the magnetic susceptibility for the z-direction when the magnetic moment operator becomes $\hat{\mu}_z = (\hat{L}_z + 2\hat{S}_z)\beta$. The calculation of the magnetic susceptibility breaks down into the following five stages:

Stage 1. A perturbation calculation is done on the four-fold degenerate level at energy $-(\lambda/2)$, which enable the first-order corrections to the energy due to the magnetic field to be calculated. Having found these we then know the $W_i^{(1)}$ coefficients in Van Vleck's equation.

Stage 2. The same calculation as in stage 1 is carried out on the level at λ.

Stage 3. The second-order corrections to the energies due to the interaction of the level at λ with that at $-(\lambda/2)$ under the action of the magnetic field are calculated, thus giving the $W_i^{(2)}$ coefficients in Van Vleck's equation.

Stage 4. Further second-order corrections to the energies due to the interaction in the magnetic field between the energy levels arising from the ^2E and $^2T_2$ levels under the action of spin–orbit coupling should be calculated. Under the present treatment however, we will assume that these are negligible compared with the energies calculated in stages 1–3, for the same reason that they were ignored in the spin–orbit coupling calculation.

Stage 5. Having found all the required Zeeman coefficients these are now substituted into Van Vleck's equation to give the magnetic susceptibilities.

The essential steps in the calculations involved in stages 1, 2, 3, and 5 will now be given.

Stage 1. Using the wave-functions (4.14), all the possible matrix elements $\langle \phi_i | \hat{L}_z + 2\hat{S}_z | \phi_j \rangle$, where i and j run from 1 to 4, are calculated and the appropriate secular determinant constructed. The evaluation of the above type of matrix element is illustrated by an example, the evaluation of the remainder is left as a useful exercise, e.g.

$$\langle \phi_2 | \hat{L}_z + 2\hat{S}_z | \phi_2 \rangle = \langle 1, -\tfrac{1}{2} | \hat{L}_z + 2\hat{S}_z | 1, -\tfrac{1}{2} \rangle$$
$$= \langle 1, -\tfrac{1}{2} | \hat{L}_z | 1, -\tfrac{1}{2} \rangle + 2 \langle 1, -\tfrac{1}{2} | \hat{S}_z | 1, -\tfrac{1}{2} \rangle$$

$$= 1\langle 1, -\tfrac{1}{2}|1, -\tfrac{1}{2}\rangle + (2)(-\tfrac{1}{2})\langle 1, -\tfrac{1}{2}|1, -\tfrac{1}{2}\rangle$$
$$= 1 - 1 = 0.$$

	$|\phi_1\rangle$	$|\phi_2\rangle$	$|\phi_3\rangle$	$|\phi_4\rangle$	
$\langle\phi_1|$	$0-E$	0	0	0	
$\langle\phi_2|$	0	$0-E$	0	0	$= 0.$ (4.16)
$\langle\phi_3|$	0	0	$0-E$	0	
$\langle\phi_4|$	0	0	0	$0-E$	

The solution to this determinant is of course $E = 0$, which tells us that to the first-order approximation the magnetic field does not remove any of the degeneracy of the level at $-(\lambda/2)$, and hence for this level $W_i^{(1)} = 0$.

Stage 2. As in stage 1 all the possible matrix elements of the type $\langle \Psi_i|\hat{L}_z + 2\hat{S}_z|\Psi_j\rangle$ are evaluated and the secular determinant constructed:

	$|\Psi_1\rangle$	$|\Psi_2\rangle$	
$\langle\Psi_1|$	$\beta H - E$	0	$= 0.$ (4.17)
$\langle\Psi_2|$	0	$-\beta H - E$	

The solutions of this determinant are $E = \pm\beta H$, i.e. the magnetic field in the first-order approximation splits the level at $E = \lambda$ into two components at energies $\pm\beta H$, which gives the first-order Zeeman coefficient for this level as $W_i^{(1)} = \pm\beta$. It is worth noting at this juncture that since the separation between adjacent Zeeman levels is defined as $g\beta H$, we have $g = 2 \cdot 00$ for the present level.

Stage 3. The second-order correction to the energies of the components of the level at $E = -(\lambda/2)$ in the magnetic field, due to the interaction between the wave-functions of the level at $E = \lambda$ and those of the level at $E = -(\lambda/2)$, takes the form

$$\sum_j \frac{\langle\phi_i|\hat{\mu}_z H|\Psi_j\rangle\langle\Psi_j|\hat{\mu}_z H|\phi_i\rangle}{E_i^0 - E_j^0}, \qquad (4.18)$$

whilst the term for the correction to the energies of the components of the level at $E = \lambda$ is

$$\sum_i \frac{\langle\Psi_j|\hat{\mu}_z H|\phi_i\rangle\langle\phi_i|\hat{\mu}_z H|\Psi_j\rangle}{E_j^0 - E_i^0}. \qquad (4.19)$$

Equation 4.18 gives the second-order correction to the energy of the ith component of the level at $E = (\lambda/2)$ and hence $W_i^{(2)}$ is known, and similarly for the jth component of the level at $E = \lambda$. However, in Van Vleck's equation all the $W_i^{(2)}$ coefficients, which belong to levels degenerate in the absence of the magnetic field, are weighted by the same exponential factor and it is convenient, for the purposes of substitution into Van Vleck's equation and also for summarizing the information on diagrams such as Fig. 4.2, to express the second-order Zeeman coefficients as a sum for all the initially degenerate components. Thus the total second-order Zeeman coefficient for the level at $E = -(\lambda/2)$ is:

$$\sum_{i=1}^{4} \sum_{j=1}^{2} \frac{\langle \phi_i | \hat{\mu}_z | \Psi_j \rangle \langle \Psi_j | \hat{\mu}_z | \phi_i \rangle}{-\dfrac{\lambda}{2} - \lambda} = \frac{-8\beta^2}{3\lambda}, \tag{4.20}$$

and for the level at $E = \lambda$

$$\sum_{j=1}^{2} \sum_{i=1}^{4} \frac{\langle \Psi_j | \hat{\mu}_z | \phi_i \rangle \langle \phi_i | \hat{\mu}_z | \Psi_j \rangle}{\lambda - \left(-\dfrac{\lambda}{2}\right)} = \frac{8\beta^2}{3\lambda}. \tag{4.21}$$

All the information calculated in stages 1–3 is summarized in Fig. 4.2.

Stage 5. The information summarized in Fig. 4.2 can now be substituted into Van Vleck's equation. It is convenient to take the level at $E = -(\lambda/2)$ as the zero for our energy scale. Hence

$$\bar{\chi}_A = N \left[\frac{\left[\dfrac{0}{kT} - 2\left(\dfrac{-8\beta^2}{3\lambda} \right) \right] \exp\left(-\dfrac{0}{kT} \right) + \left[\dfrac{\beta^2}{kT} + \dfrac{(-\beta)^2}{kT} - 2\left(\dfrac{8\beta^2}{3\lambda} \right) \right] \exp\left(-\dfrac{3\lambda}{2kT} \right)}{4 \exp\left(-\dfrac{0}{kT} \right) + 2 \exp\left(-\dfrac{3\lambda}{2kT} \right)} \right]$$

$$\bar{\chi}_A = \frac{N\beta^2}{3kT} \frac{\left[\dfrac{16kT}{\lambda} + \left(6 - \dfrac{16kT}{\lambda} \right) \exp\left(-\dfrac{3\lambda}{2kT} \right) \right]}{\left[4 + 2 \exp\left(-\dfrac{3\lambda}{2kT} \right) \right]}.$$

Substituting $x = \lambda/kT$

$$\bar{\chi}_A = \frac{N\beta^2}{3kT} \frac{\left[8 + (3x - 8) \exp\left(-\dfrac{3x}{2} \right) \right]}{x \left[2 + \exp\left(-\dfrac{3x}{2} \right) \right]},$$

or

$$\bar{\mu}^2 = \frac{\left[8 + (3x - 8) \exp\left(-\frac{3x}{2}\right)\right]}{x\left[2 + \exp\left(-\frac{3x}{2}\right)\right]}.$$ (4.22)

The variation of $\bar{\mu}$ with temperature given by Equation 4.22 is illustrated in Fig. 4.3(a) (in order to plot the behaviour on a convenient scale, the temperature is expressed in terms of the parameter kT/λ). The more detailed variations of the magnetic moment and susceptibilities expected for the specific case of octa-hedral titanium(III), assuming the free ion value of the spin–orbit coupling constant, are given in Figs. 4.3(b) and 4.3(c). At room temperature the expected magnetic moment is about 1·88 B.M. and this falls with temperature approach-ing zero at 0 K. At temperature below about 100 K, $\bar{\mu}$ varies approximately as $T^{1/2}$ because the *susceptibility* becomes independent of temperature (see Fig. 4.3(c)). The variation of $1/\chi_{Ti}$ with temperature shows another interesting feature in that above about 200 K a Curie–Weiss law is obeyed with $\theta = 200°$. This feature is important in that large values of θ are often associated with the presence of antiferromagnetic interactions (for further discussion of this see Chapter 7), but in this system this is not the case, and it must be emphasized that such a practice is dangerous unless other information, such as the occurrence of a maximum in the susceptibility, is available.

A good deal of useful qualitative information concerning the behaviour of the magnetic susceptibility of an octahedral titanium(III) complex can be deduced simply by examining the splitting-diagram Fig. 4.2. The lowest energy level after the action of spin–orbit coupling is four-fold degenerate and is not split or changed in energy by the magnetic field in the first-order approximation. Since the magnetic susceptibility depends on the rate of change of energy in the magnetic field (see Chapter 1), then to the first order of approximation this lowest level has zero susceptibility. The only contribution to the susceptibility due to this level arises from the second-order Zeeman effect and thus at low temperatures where only this lowest level is thermally populated the *sus-ceptibility* will be independent of temperature. As the temperature is raised the thermal energy becomes comparable with the separation between the spin–orbit levels leading to increasing population of the higher level. Since this level has a first-order Zeeman contribution the *susceptibility* now becomes temperature dependent.

Tetrahedral copper(II) provides an example of a complex with a $^2T_2$ ground term for which λ is negative. Because λ is negative the splitting diagram (Fig. 4.2) will be inverted, and this leads to the variation of $\bar{\mu}$ with temperature as

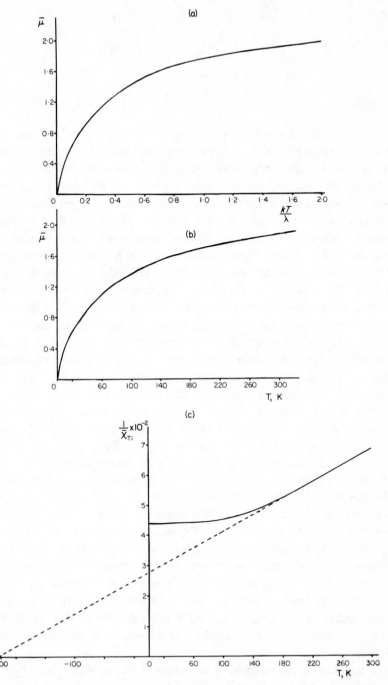

Figure 4.3. The variation of (a) $\bar{\mu}$ versus $kT/|\lambda|$ for the $^2T_2$ term, (b) $\bar{\mu}$ versus T for octahedral titanium(III), (c) $1/\bar{\chi}_{Ti}$ *vs.* T for octahedral titanium(III). In (b) and (c) the free-ion value of λ is used.

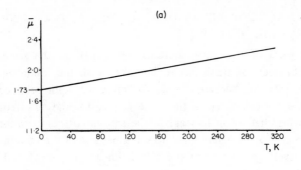

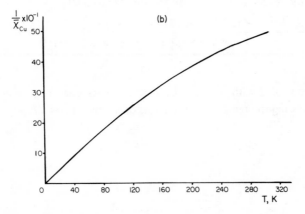

Figure 4.4. The variation of (a) $\bar{\mu}$ versus T and (b) $1/\bar{\chi}_{Cu}$ *vs.* T, for tetrahedral copper(II) assuming the free ion value of λ.

shown in Fig. 4.4(a). The variation of the inverse of the susceptibility with temperature, Fig. 4.4(b), corresponds essentially to a Curie-law behaviour with a temperature independent contribution, hence the curvature. The general features of the above variations can be qualitatively deduced from the splitting diagram as we did in the case of titanium(III). The lowest level is split into two components by the magnetic field and has a g value of $2 \cdot 00$, as expected when there is no orbital contribution associated with a single electron. Thus in the limit of zero temperature we may expect the magnetic moment to correspond to the spin-only value for one unpaired electron. As the temperature is raised the magnetic susceptibility will be that for one unpaired electron plus a temperature independent contribution arising from the second-order Zeeman effect. This will continue until a stage is reached where the upper 'non-magnetic' level is thermally populated, which results in the susceptibility being that due to the spin-only value for one unpaired electron plus the second-order Zeeman contributions. These will not necessarily give rise to T.I.P. since they are weighted by $\exp\left(-W_j^0/kT\right)$, which will no longer approximate to zero.

4.3.2. THE MAGNETIC BEHAVIOUR OF THE $^5T_2$ TERM WHICH ARISES FROM A FREE ION ^5D TERM [1]

The orbital parts of the wave-functions (i.e. $|M_L\rangle$) are the same as for the $^2T_2$ term but each of these has to be taken with the following spin parts, $|M_S\rangle = |\pm 2\rangle, |\pm 1\rangle, |0\rangle$, because $S = 2$. The effects of spin–orbit coupling and magnetic field are calculated in the same way as for the $^2T_2$ term, the only difference being that of an increase in the amount of computation, since we initially have a 15 × 15 determinant to solve, because of the 15-fold degeneracy of the $^5T_2$ level. The details of the calculation are not given and we merely summarize the results in Fig. 4.5 and the following equation.

$$\bar{\mu}^2 = \frac{3[28x + 9\cdot33 + (22\cdot5x + 4\cdot17)\exp(-3x) + (24\cdot5x - 13\cdot5)\exp(-5x)]}{x[7 + 5\exp(-3x) + 3\exp(-5x)]}.$$

(4.23)

The behaviour of $\bar{\mu}$ with temperature is summarized for the general case in Fig. 4.6 by expressing the temperature in terms of the parameter $kT/|\lambda|$.

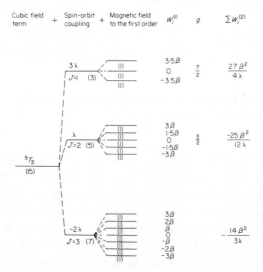

Figure 4.5. The splitting diagram for the $^5T_2$ term under the action of spin–orbit coupling and a magnetic field.

The most common example of a transition metal ion which has a $^5T_2$ ground term is octahedral spin-free iron(II), for which λ is negative. The variation of the magnetic moment and the reciprocal of the susceptibility with temperature are given in Figs. 4.7(a) and (b) assuming the free ion value for the spin–orbit coupling constant, i.e. $\lambda = -100$ cm^{-1}.

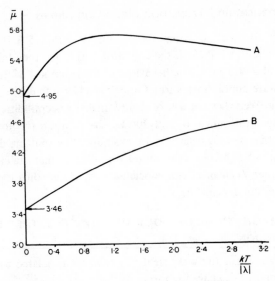

Figure 4.6. The variation of $\bar{\mu}$ versus $kT/|\lambda|$ for the $^5T_2$ term when λ is (A) negative and (B) positive.

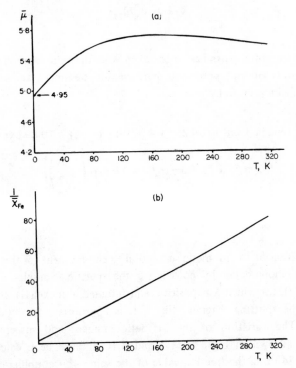

Figure 4.7. The variation of (a) $\bar{\mu}$ versus T and (b) $1/\bar{\chi}_{Fe}$ *vs.* T, for octahedral iron(III) assuming the free ion value of λ.

The magnetic moment at room temperature is about 5·6 B.M. rising to a maximum of about 5·7 B.M. in the region of 160 K and then falling steadily to 4·95 B.M. at absolute zero. The behaviour of the reciprocal of the susceptibility with temperature approximates to a Curie–Weiss law with $\theta \approx 6°$. The deviation from the Curie–Weiss law consists of slightly higher susceptibilities than required in the temperature range of 60–290 K. It is not easy to predict, even qualitatively, this type of behaviour by examining the splitting diagram for the $^5T_2$ term, Fig. 4.5. The best one can do is to observe that each level has a first- and second-order Zeeman effect associated with it and thus the *susceptibility* will be temperature dependent.

4.3.3. THE MAGNETIC BEHAVIOUR OF THE $^3T_1$ TERM WHICH ARISES FROM A FREE ION ^3F TERM [1]

The wave-functions for the weak field $^3T_1$ term are associated with $L = 3$ and $S = 1$ and in terms of $|M_L\rangle$ have the form

$$\sqrt{\tfrac{5}{8}}|-3\rangle + \sqrt{\tfrac{3}{8}}|1\rangle$$
$$\sqrt{\tfrac{5}{8}}|3\rangle + \sqrt{\tfrac{3}{8}}|1\rangle$$
$$|0\rangle$$

Each of these orbital parts has to be taken with the spin components $|\pm 1\rangle$ and $|0\rangle$. The results of the appropriate perturbation treatment are summarized in Fig. 4.8, whilst $\bar{\mu}$ is given by

$$\bar{\mu}^2 = \frac{3\left[0·625x + 6·8 + (0·125x + 4·09)\exp(-3x) - 10·89\exp\left(-\dfrac{9x}{2}\right)\right]}{x\left[5 + 3\exp(-3x) + \exp\left(-\dfrac{9x}{2}\right)\right]}.$$

$$(4.24)$$

The behaviour of $\bar{\mu}$ with temperature is summarized for the general case in Fig. 4.9.

The variation of the magnetic susceptibility and moment for this term will be discussed in more detail by considering the specific examples of octahedral vanadium(III), for which λ is positive, and tetrahedral nickel(II), for which λ is negative. The splitting diagram, Fig. 4.8, is appropriate to octahedral vanadium(III). The variation of the magnetic moment and susceptibility with temperature shown in Figs. 4.10(a) and 4.10(b) have been calculated from Equation 4.24 using the free ion value of the spin–orbit coupling constant, $\lambda = 105$ cm^{-1}. The magnetic moment expected for octahedral vanadium(III) at room temperature is about 2·7 B.M. and this is expected to fall with temperature until a value of 0·62 B.M. is reached at 0 K. It is noteworthy that the magnetic susceptibility, which gives rise to the above variation in $\bar{\mu}$, follows a Curie–Weiss law, in the temperature range 320–80 K, with a θ value of several hundred

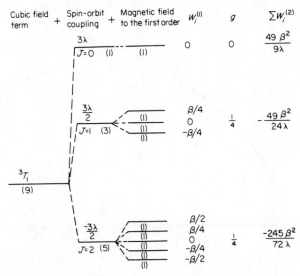

Figure 4.8. The splitting diagram for the weak field $^3T_1$ term under the action of spin-orbit coupling and magnetic field.

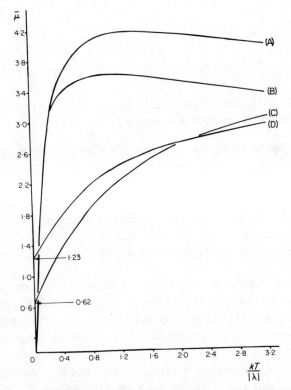

Figure 4.9. The variation of $\bar{\mu}$ versus $kT/|\lambda|$ for the $^3T_1$ term, when (A) λ is negative, A = 1·5 (weak field); (B) λ is negative, A = 1·0 (strong field); (C) λ is positive, A = 1·5 (weak field); (D) λ is positive, A = 1·0 (strong field).

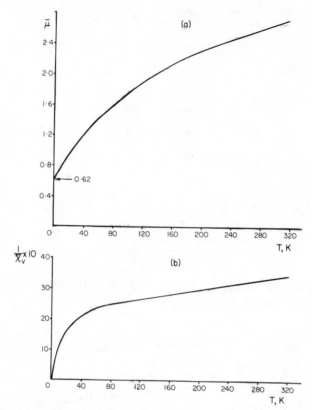

Figure 4.10. The variation of (a) $\bar{\mu}$ *vs.* T and (b) $1/\bar{X}_V$ *vs.* T, for octahedral vanadium(III) assuming the free ion value of λ.

degrees, again demonstrating that large values of θ do not necessarily have their origins in antiferromagnetic exchange.

Examination of the splitting diagram does not enable any very definite conclusions to be drawn concerning the behaviour of the magnetic susceptibility with temperature. The main qualitative deduction which can be made is that, since the lower levels both have first- and second-order Zeeman effects associated with them, we would expect the *susceptibility* to be always temperature dependent.

In the case of tetrahedral nickel(II), λ is negative and consequently the splitting diagram will be inverted. The change in the sign of λ also changes the sign of the various second order Zeeman coefficients $\Sigma W_i^{(2)}$. The variation of the magnetic moment and susceptibility with temperature assuming the free ion value of $\lambda = -315$ cm^{-1} is shown in Figs. 4.11(a) and 4.11(b), respectively. The value of $\bar{\mu}$ at room temperature is about 4.0 B.M., and falls with temperature until it becomes zero at 0 K. Below about 120 K $\bar{\mu}$ varies as $T^{1/2}$, since in this

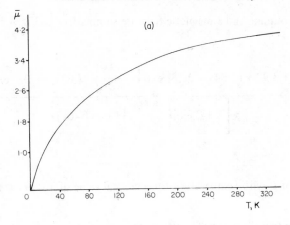

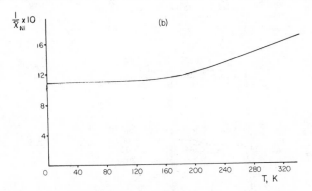

Figure 4.11. The variation of (a) $\bar{\mu}$ *vs.* T and (b) $1/\bar{\chi}_{Ni}$ *vs.* T, for tetrahedral nickel(II) assuming the free ion value of λ.

region the magnetic *susceptibility* is independent of temperature. Above about 200 K a Curie–Weiss law is obeyed with a θ value of about 160°. The information on the splitting diagram enables us to predict that at very low temperatures the magnetic *susceptibility* will be independent of temperature, since the lowest energy level has only a second-order Zeeman effect associated with it. As the temperature is raised the levels for which there are both first- and second-order Zeeman effects become thermally populated, and the magnetic *susceptibility* becomes temperature dependent.

4.3.4. THE MAGNETIC BEHAVIOUR OF THE $^4T_1$ TERM WHICH ARISES FROM A FREE ION ^4F TERM [1]

The orbital parts of the wave-functions for the weak field $^4T_1$ term are same as those for the $^3T_1$ term. Each of these orbital parts has now to be taken with the spin components $|\pm\frac{3}{2}\rangle$, $|\pm\frac{1}{2}\rangle$, since $S = \frac{3}{2}$. The results of consecutive perturbations by spin–orbit coupling and a magnetic field are summarized in Fig. 4.12, and $\bar{\mu}$ is given by:

$$\bar{\mu}^2 = \frac{3\left[3 \cdot 15x + 3 \cdot 92 + (2 \cdot 84x + 2 \cdot 13)\exp\left(-\dfrac{15x}{4}\right) + (4 \cdot 7x - 6 \cdot 05)\exp\left(-6x\right)\right]}{x\left[3 + 2\exp\left(-\dfrac{15x}{4}\right) + \exp\left(-6x\right)\right]}.$$

(4.25)

The variation of $\bar{\mu}$ with temperature for a general value of λ is illustrated in Fig. 4.13.

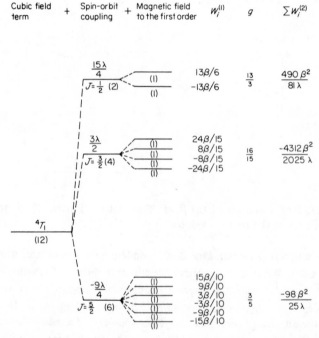

Figure 4.12. The splitting diagram for the weak field $^4T_1$ term under the action of spin–orbit coupling and a magnetic field.

The most common example of a transition metal which gives rise to a $^4T_1$ ground term is octahedral cobalt(II) and its magnetic behaviour will be examined in more detail. The splitting diagram, Fig. 4.12, has to be inverted for octahedral cobalt(II) because we have a d-shell more than half full and λ is negative. The resulting variation of the magnetic moment and susceptibility with temperature is shown in Figs. 4.14(a) and 4.14(b), assuming the free ion value for the spin–orbit coupling constant, $\lambda = -170$ cm^{-1}. At room temperature $\bar{\mu}$ is

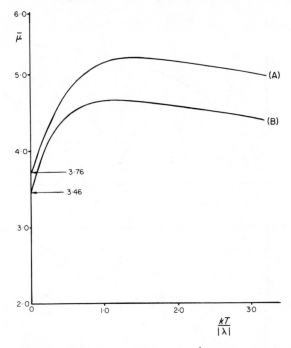

Figure 4.13. The variation of $\bar{\mu}$ *vs.* $kT/|\lambda|$ for the $^4T_1$ term with λ negative when (A) A = 1·5 (weak field) and (B) A = 1·0 (strong field).

expected to be about 5.2 B.M. and it will fall steadily, but not linearly, until a value of 3.76 B.M. is reached at 0 K. The magnetic susceptibility obeys a Curie–Weiss law with $\theta \approx 20°$ over the temperature range 50–320 K. Below 50 K the susceptibility increases more rapidly than required by this Curie–Weiss law and $1/\bar{\chi}_{Co}$ tends towards zero at 0 K. Examination of the splitting diagram does not enable any very detailed predictions to be made, except to note that, because all of the spin–orbit levels have both first- and second-order Zeeman effects associated with them, we would expect the susceptibility to always be temperature dependent.

4.3.5. THE MAGNETIC BEHAVIOUR OF A_2 TERMS WHICH ARISE FROM FREE ION F TERMS [1]

In certain instances a cubic crystal field may split the free ion F term in such a way that an orbitally non-degenerate A_2 term lies lowest, see Table 4.2. Since there is no orbital angular momentum associated with this term, there is nothing for the spin–angular momentum to couple with, and thus spin–orbit coupling cannot raise any of the degeneracy. Another consequence of this lack of orbital angular momentum is that we ought to observe the spin-only values for the magnetic susceptibilities. However, as we will see shortly, from a more detailed

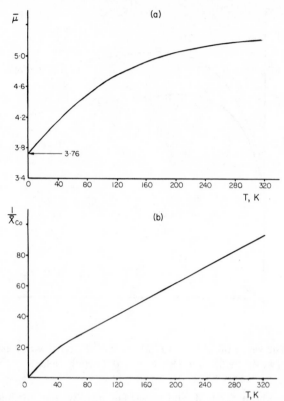

Figure 4.14. The variation of (a) $\bar{\mu}$ *vs.* T and (b) $1/\bar{\chi}_{Co}$ *vs.* T, for octahedral cobalt(II) assuming the free ion value of λ.

calculation of the $^3A_2$ ground term, spin–orbit coupling 'mixes in' some of a higher-lying orbital triplet term which effectively introduces an orbital contribution into our initially orbitally non-degenerate ground term.

The calculation of the magnetic susceptibility of the $^3A_2$ term which arises from the ^3F term.

Although spin–orbit coupling does not split the $^3A_2$ term, there are non-zero matrix elements, with the higher-lying ^3T terms which arise from the free ion ^3F term. If we take our perturbation calculation to the second-order of approximation we find that spin–orbit coupling modifies the wave-functions of the $^3A_2$ term by mixing small amounts of the wave-functions of the higher ^3T terms. This 'mixing-in' of the wave-functions has the effect of introducing a small orbital contribution into the susceptibility. Whereas in the cases of the orbital triplet ground states we ignored this second-order mixing due to spin–orbit coupling, it is included here because its effect should be more noticeable in the absence of any first-order effect.

The magnetic susceptibility of the $^3A_2$ term will now be calculated under the assumption that only the 'mixing-in' of the $^3T_2$ term, which is $10Dq$ above the ground term, is important. The wave-functions for the $^3A_2$ term, written as $|M_L, M_S\rangle$ are:

$$\left.\begin{aligned}
\phi_1 &= \sqrt{\tfrac{1}{2}}(|2, 1\rangle - |-2, 1\rangle) \\
\phi_2 &= \sqrt{\tfrac{1}{2}}(|2, 0\rangle - |-2, 0\rangle) \\
\phi_3 &= \sqrt{\tfrac{1}{2}}(|2, -1\rangle - |-2, -1\rangle)
\end{aligned}\right\} \qquad (4.26)$$

whilst for the $^3T_2$ term they are:

$$\left.\begin{aligned}
\phi_4 &= \sqrt{\tfrac{5}{8}}|1, 1\rangle - \sqrt{\tfrac{3}{8}}|-3, 1\rangle \\
\phi_5 &= \sqrt{\tfrac{5}{8}}|1, 0\rangle - \sqrt{\tfrac{3}{8}}|-3, 0\rangle \\
\phi_6 &= \sqrt{\tfrac{5}{8}}|1, -1\rangle - \sqrt{\tfrac{3}{8}}|-3, -1\rangle \\
\phi_7 &= \sqrt{\tfrac{5}{8}}|-1, 1\rangle - \sqrt{\tfrac{3}{8}}|3, 1\rangle \\
\phi_8 &= \sqrt{\tfrac{5}{8}}|-1, 0\rangle - \sqrt{\tfrac{3}{8}}|3, 0\rangle \\
\phi_9 &= \sqrt{\tfrac{5}{8}}|-1, -1\rangle - \sqrt{\tfrac{3}{8}}|3, -1\rangle \\
\phi_{10} &= \sqrt{\tfrac{1}{2}}(|2, 1\rangle + |-2, 1\rangle) \\
\phi_{11} &= \sqrt{\tfrac{1}{2}}(|2, 0\rangle + |-2, 0\rangle) \\
\phi_{12} &= \sqrt{\tfrac{1}{2}}(|2, -1\rangle + |-2, -1\rangle)
\end{aligned}\right\} . \qquad (4.27)$$

In order to calculate the extent of 'mixing in' that occurs we need to calculate matrix elements of the type $\langle\phi_j|\lambda\hat{L} . \hat{S}|\phi_i\rangle$ where $i = 1$ to 3 and $j = 4$ to 12, see Chapter 3. The only non-zero matrix elements which occur are:

$$\langle\phi_8|\lambda\hat{L} . \hat{S}|\phi_1\rangle = -2\lambda, \qquad\qquad \langle\phi_{10}|\lambda\hat{L} . \hat{S}|\phi_1\rangle = 2\lambda$$

$$\langle\phi_4|\lambda\hat{L} . \hat{S}|\phi_2\rangle = 2\lambda, \qquad\qquad \langle\phi_9|\lambda\hat{L} . \hat{S}|\phi_2\rangle = -2\lambda$$

$$\langle\phi_5|\lambda\hat{L} . \hat{S}|\phi_3\rangle = 2\lambda, \qquad\qquad \langle\phi_{12}|\lambda\hat{L} . \hat{S}|\phi_3\rangle = -2\lambda$$

By defining the wave-functions as above, the coefficients, c_k, in Equation 3.25 are equal to unity. The energy separation $E_i - E_j = -10Dq$ and providing terms in $(\lambda/10Dq)^2$ are negligible compared to unity in the normalizing coefficient, the modified wave-functions for the $^3A_2$ ground term become:

$$\left.\begin{aligned}
\psi_1 &= \phi_1 - \frac{2\lambda}{10Dq}(\phi_{10} - \phi_8) \\
\psi_2 &= \phi_2 - \frac{2\lambda}{10Dq}(\phi_4 - \phi_9) \\
\psi_3 &= \phi_3 - \frac{2\lambda}{10Dq}(\phi_5 - \phi_{12})
\end{aligned}\right\} \qquad (4.28)$$

It is worth observing at this point that 'mixing in' due to spin–orbit coupling does not raise the degeneracy of the $^3A_2$ term. However, the energies of all the components do not remain unchanged, but are each depressed by an amount

$$\sum_{j=4}^{12} \frac{\langle \phi_i | \lambda \hat{L} \cdot \hat{S} | \phi_j \rangle \langle \phi_j | \lambda \hat{L} \cdot \hat{S} | \phi_i \rangle}{E_i - E_j} = -\frac{8\lambda^2}{10Dq}.$$

The magnetic field may now be treated as a further perturbation on the system specified by ψ_1, ψ_2, and ψ_3. The energy levels in the magnetic field, in the first-order approximation, are obtained from the following secular determinant, where terms in $(\lambda/10Dq)^2$ have been neglected.

$$\begin{array}{c|ccc}
 & |\psi_1\rangle & |\psi_2\rangle & |\psi_3\rangle \\
\hline
\langle\psi_1| & \left(2 - \dfrac{8\lambda}{10Dq}\right)\beta H - E & 0 & 0 \\
\langle\psi_2| & 0 & 0 - E & 0 \\
\langle\psi_3| & 0 & 0 & \left(-2 + \dfrac{8\lambda}{10Dq}\right)\beta H - E
\end{array} = 0. \quad (4.29)$$

The solutions to this determinant are:

$$E = \pm\left(2 - \frac{8\lambda}{10Dq}\right)\beta H \text{ and } 0. \quad (4.30)$$

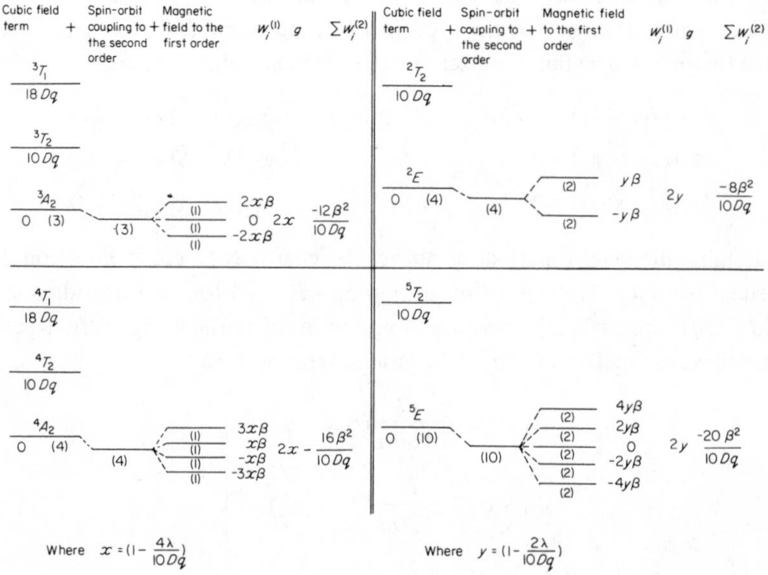

Figure 4.15. The splitting diagrams for A_2 and E ground terms under the action of spin–orbit coupling and a magnetic field.

In the first-order approximation the magnetic field has split the $^3A_2$ term into three components, as shown in Fig. 4.15, the appropriate g value being $2[1 - (4\lambda/10Dq)]$. In addition to this first-order Zeeman effect there is a second-order Zeeman effect due to 'mixing' in the magnetic field between the wave-functions of the components of the $^3T_2$ term. The total second-order Zeeman effect of the components of the $^3T_2$ term on the $^3A_2$ term is given by the expression

$$\sum_{i=1}^{3} \sum_{j=4}^{12} \frac{\langle \psi_i|(\hat{L}_z + 2\hat{S}_z)\beta H|\phi_j\rangle\langle\phi_j|(\hat{L}_z + 2\hat{S}_z)\beta H|\psi_i\rangle}{E_i - E_j}. \tag{4.31}$$

If terms in $\lambda/10Dq$ in the matrix elements, and the energy shifts due to spin–orbit coupling are assumed to be small so that $E_i - E_j = -10Dq$, the total second-order Zeeman coefficient, $\sum_i W_i^{(2)}$, for the $^3A_2$ term becomes $-[(3)(4)\beta^2/10Dq]$. Substitution of the calculated first- and second-order Zeeman coefficients into Van Vleck's equation gives the susceptibility

$$\bar{\chi}_A = \frac{N\beta^2}{3kT} 2\left(2 - \frac{8\lambda}{10Dq}\right)^2 + \frac{8N\beta^2}{10Dq}. \tag{4.32}$$

If the T.I.P. contribution is subtracted from the susceptibility, the expression for the magnetic moment is

$$\bar{\mu} = \sqrt{8}\left(1 - \frac{4\lambda}{10Dq}\right) = \mu_{s.o.}\left(1 - \frac{4\lambda}{10Dq}\right). \tag{4.33}$$

where $\mu_{s.o} = (4S(S + 1))^{1/2}$ = the spin-only magnetic moment. The magnetic susceptibility given by Equation 4.32 represents Curie-law behaviour with a superimposed T.I.P., whilst $\bar{\mu}$ will be independent of temperature providing allowance is made for the T.I.P. contribution to the susceptibility.

The magnetic behaviour of the $^4A_2$ term which arises from the free ion ^4F term [1]

The magnetic properties of the $^4A_2$ term which arises from the free ion ^4F term are analogous to those of the $^3A_2$ term we have just discussed. The corresponding expressions for the magnetic susceptibilities and moments are:

$$\bar{\chi}_A = \frac{N\beta^2}{3kT} 15\left(1 - \frac{4\lambda}{10Dq}\right)^2 + \frac{8N\beta^2}{10Dq}, \tag{4.34}$$

and

$$\bar{\mu} = \sqrt{15}\left(1 - \frac{4\lambda}{10Dq}\right) = \mu_{s.o.}\left(1 - \frac{4\lambda}{10Dq}\right). \tag{4.35}$$

The relevant data for the magnetic behaviour of all the A and E ground terms for octahedral and tetrahedral transition metal complexes is summarized in Fig. 4.15 and Table 4.2.

Table 4.2. *The occurrence of A and E terms for spin-free d^n configurations and the relationship between $\bar{\mu}$ and $\mu_{s.o.}$ for these terms.*

No. of d-electrons	Stereo-chemistry	Ground term	Sign of λ	α in $\bar{\mu} = \mu_{s.o.}\left(1 - \dfrac{\alpha\lambda}{10Dq}\right)$	$\bar{\mu}$ compared with $\mu_{s.o.}$
1	Tetrahedral	^2E	positive	2	less than
2	Tetrahedral	$^3A_2$	positive	4	less than
3	Octahedral	$^4A_{2g}$	positive	4	less than
4	Octahedral	^5E_g	positive	2	less than
5	Octahedral	$^6A_{1g}$	−	0	equal to
5	Tetrahedral	$^6A_1$	−	0	equal to
6	Tetrahedral	^5E	negative	2	greater than
7	Tetrahedral	$^4A_2$	negative	4	greater than
8	Octahedral	$^3A_{2g}$	negative	4	greater than
9	Octahedral	^2E_g	negative	2	greater than

4.3.6. THE MAGNETIC BEHAVIOUR OF E TERMS WHICH ARISE FROM FREE ION D TERMS [1]

The orbital wave-functions for the ^2E and ^5E terms, which can be ground terms for transition metal complexes are $|0\rangle$ and $\sqrt{\frac{1}{2}}(|2\rangle + |-2\rangle)$. Since $\langle\psi|\hat{L}_z|\psi\rangle = 0$, when $\psi = |0\rangle$ or $\sqrt{\frac{1}{2}}(|2\rangle + |-2\rangle)$, and $\hat{L}_+$ and $\hat{L}_-$ can only connect wave-functions whose M_L values differ by ± 1, there is no orbital angular momentum associated with these E terms and hence spin–orbit coupling will not raise any of their degeneracy. As in the case of the A_2 ground terms this lack of orbital angular momentum should lead to the spin-only value for magnetic moment in the first approximation. However, we can again get mixing under the action of spin–orbit coupling between the E ground term and the higher lying T_2 term. When this mixing is taken into account we have the following expression for the magnetic susceptibility and moment. For the ^2E term:

$$\bar{\chi}_A = \frac{N\beta^2}{3kT} \, 3 \left(1 - \frac{2\lambda}{10Dq}\right)^2 + \frac{4N\beta^2}{10Dq}, \qquad (4.36)$$

and after allowing for the T.I.P. term:

$$\bar{\mu} = \sqrt{3}\left(1 - \frac{2\lambda}{10Dq}\right) = \mu_{s.o.}\left(1 - \frac{2\lambda}{10Dq}\right). \qquad (4.37)$$

For the ^5E term:

$$\bar{\chi}_A = \frac{N\beta^2}{3kT} 24 \left(1 - \frac{2\lambda}{10Dq}\right) + \frac{4N\beta^2}{10Dq}, \tag{4.38}$$

and as above:

$$\bar{\mu} = \sqrt{24} \left(1 - \frac{2\lambda}{10Dq}\right) = \mu_{s.o.} \left(1 - \frac{2\lambda}{10Dq}\right). \tag{4.39}$$

Table 4.3. *The values of $\bar{\mu}$ for some typical transition metal complexes with A and E ground terms.*

No. of d-electrons	Compound	Stereo-chemistry	Ground term	$\mu_{s.o.}$ B.M.	$\bar{\mu}$ at 300 K	$\bar{\mu}$ at 80 K
1	VCl$_4$	Tetrahedral	^2E	1·73	1·72	—
3	KCr(SO$_4$)$_2$12H$_2$O	Octahedral	$^4A_{2g}$	3·87	3·84	3·84
	K$_3$Cr(CN)$_6$	Octahedral	$^4A_{2g}$	3·87	3·87	3·87
	K$_3$MoCl$_6$	Octahedral	$^4A_{2g}$	3·87	3·79	3·62
	K$_2$ReCl$_6$	Octahedral	$^4A_{2g}$	3·87	3·25	2·57
4	Cr(SO$_4$)6H$_2$O	Octahedral	^5E_g	4·90	4·82	4·84
	Mn(acetylacetonate)$_3$	Octahedral	^5E_g	4·90	4·86	4·75
5	K$_2$Mn(SO$_4$)$_2$6H$_2$O	Octahedral	$^6A_{1g}$	5·92	5·92	5·92
	KFe(SO$_4$)$_2$12H$_2$O	Octahedral	$^6A_{1g}$	5·92	5·89	5·89
	(Et$_4$N)$_2$MnCl$_4$	Tetrahedral	$^6A_1$	5·92	5·94	—
	(Et$_4$N)FeCl$_4$	Tetrahedral	$^6A_1$	5·92	5·88	—
6	(Et$_4$N)$_2$FeCl$_4$	Tetrahedral	^5E	4·90	5·40	5·40
7	Cs$_2$CoCl$_4$	Tetrahedral	$^4A_2$	3·87	4·71	4·48
8	(NH$_4$)$_2$Ni(SO$_4$)$_2$6H$_2$O	Octahedral	$^3A_{2g}$	2·83	3·23	3·20
9	K$_2$Cu(SO$_4$)$_2$6H$_2$O	Octahedral	^2E_g	1·73	1·91	1·91

The magnetic properties of transition metal ions with A_2 or E ground terms obviously depend on the values of λ and $10Dq$. When the metal ion in a complex has less than a half-filled d-shell, λ is positive and we would expect $\bar{\mu}$ to be less than the spin-only value. Reference to the examples in Table 4.3 confirm this, although in general because of the small values of λ compared with $10Dq$ for these complexes, the expected decrease in $\bar{\mu}$ is small. On the other hand, when the d-shell is more than half full, λ is negative and $\bar{\mu}$ greater than the spin-only value is expected. This prediction is borne out in practice and particularly favourable examples are normally found for six-coordinate nickel(II) and copper(II). The variation of the magnetic susceptibility with temperature of complexes with A_2 or E ground terms is illustrated in Fig. 4.16 with the example of $[Ni(H_2O)_6]^{2+}$, using the free ion value of λ and $10Dq = 9\,000$ cm^{-1}. The plot of $1/\bar{\chi}_{Ni}$ against temperature follows a Curie-law up to about

150 K, but above this temperature there is a pronounced curvature. This curvature represents the effects of the T.I.P. contribution which becomes more noticeable as the temperature dependent contribution decreases with increasing temperature.

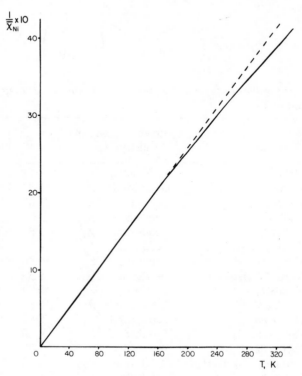

Figure 4.16. The variation of $1/\bar{\chi}_{Ni}$ *vs.* T for octahedral nickel(II) assuming $10Dq = 9\,000$ cm^{-1} and λ equal to the free ion value.

4.3.7. THE MAGNETIC BEHAVIOUR OF THE $^6A_1$ TERM WHICH ARISES FROM THE FREE ION ^6S TERM

The ^6S term which arises from the spin-free d^5 configuration is not split by crystal fields of any symmetry, but gives rise to a $^6A_1$ term. This term is an orbital singlet and consequently it has no orbital angular momentum associated with it. Thus spin–orbit coupling cannot raise the degeneracy of this term, and furthermore because there are no excited terms with the same multiplicity as the ground term there can be no second-order mixing due to spin–orbit coupling nor can there be any second-order Zeeman effect. The magnetic properties are therefore simply due to the spin contribution and

$$\bar{\mu} = (4S(S + 1))^{\frac{1}{2}} \text{B.M.} = 5 \cdot 92 \text{ B.M.}$$

which is independent of temperature. The best example of complexes with the $^6A_1$ ground term are provided by spin-free octahedral or tetrahedral manganese(II) or iron(III), and some typical complexes and their magnetic moments are given in Table 4.3.

4.4. The magnetic properties of transition metal ions in medium and strong octahedral and tetrahedral crystal fields

In Chapter 2 it was shown that, for d-electron configurations which gave a free ion F term lowest, there is also at $15B$ above it a P term of the same multiplicity. In a cubic crystal field these two terms both give rise to T_1 terms and as the strength of the crystal field increases the wave-functions of these two terms are increasingly mixed. The consequence of this is that the magnetic properties of the T_1 ground terms are modified compared to the weak field case. In contrast to this, the magnetic properties of the cubic field T_2 terms, which arise from free ion D terms are unaltered, providing of course there is no change in spin multiplicity, due to the increased crystal field strength. The reason that the magnetic properties of these latter terms remain unaltered when the crystal field strength is increased is that there are no cubic field T_2 terms of the same multiplicity as the ground term which can be mixed into the ground term by the action of the cubic crystal field. For the same reasons the magnetic properties of the A_2 or E ground terms arising from free ion F and D terms are unaltered by the crystal field strength

4.4.1. THE MAGNETIC PROPERTIES OF MEDIUM CRYSTAL FIELD T_1 TERMS WHICH ARISE FROM FREE ION F TERMS

In the present context, a medium crystal field is such that there has been no change in the spin-multiplicity of the ground term compared with the weak field case. As indicated in Chapter 2, the wave-functions for the T_1 ground term can be expressed as linear combinations of those for the $T_1(F)$ and $T_1(P)$ terms:

$$\Psi(T_1) = (1 + c_i^2)^{-\frac{1}{2}}[\Psi(T_1^0(F)) + c_i\Psi(T_1^0(P))]. \tag{4.40}$$

The mixing coefficient $c_i = (6 + E)/4$, where E is the lowest root of the equation:

$$\left. \begin{array}{c} E^2 + (6 - Z)E - 6Z - 16 = 0 \\[2em] Z = \dfrac{E(P) - E(F)}{Dq} \end{array} \right\} \tag{4.41}$$

and

The coefficient, c_i, can vary from $c_i = 0$ for a weak ligand field, to $c_i = -\frac{1}{2}$ for a strong ligand field.

Following the procedures in previous sections we will discuss the magnetic properties of the T_1 terms neglecting any contributions which may arise from any other excited terms, except of course the P terms. This can conveniently be done by defining a parameter, $A = (1\cdot5 - c_i^2)/(1 + c_i^2)$, and expressing the orbital parts of the wave-functions of the T_1 terms as combinations of $|L, M_L\rangle$ as follows [2]:

$$|\pm A\rangle = (1 + c_i^2)^{-\frac{1}{2}}[\sqrt{\tfrac{15}{24}}|3, \pm3\rangle - \sqrt{\tfrac{9}{24}}|3, \mp1\rangle + c_i|1, \mp1\rangle],$$
$$|0\rangle = (1 + c_i^2)^{-\frac{1}{2}}[|3, 0\rangle + c_i|1, 0\rangle]. \tag{4.42}$$

The orbitals defined in this way then have the property that the only non-zero matrix elements with the orbital angular momentum operators are:

$$\left.\begin{array}{c} \langle\pm A|\hat{L}_z|\pm A\rangle = \pm A \\ \langle\pm A|\hat{L}_x|0\rangle = -\sqrt{\tfrac{3}{2}}A \\ \langle\pm A|\hat{L}_y|0\rangle = -i\sqrt{\tfrac{3}{2}}A \end{array}\right\}. \tag{4.43}$$

The verification of these matrix elements provides a useful and instructive exercise for the reader. The parameter A gives a measure of the crystal field strength relative to the interelectronic repulsions, and is equal to 1·5 for a weak crystal field and 1·0 for a strong field.

$^3T_1$ terms

The result of spin–orbit coupling is to split this term into three components at energies $-A\lambda$, $A\lambda$, and $2A\lambda$; see Fig. 4.17. When $A = 1\cdot5$ we have, as expected, the same result as calculated previously for the weak crystal field case, and when $A = 1\cdot0$ the result is that for the corresponding $t_2^n e^m$ configuration to be discussed in Section 4.5. The calculation of the effects of spin–orbit coupling and a magnetic field using the perturbation method are summarized in Fig. 4.17 and lead to

$$\bar{\mu}^2 = \cfrac{\left[\begin{array}{c} 2\cdot5(A-2)^2x + \dfrac{5(A+2)^2}{6A} + \left\{0\cdot5(A-2)^2x + \dfrac{(A+2)^2}{2A}\right\} \\ \times \exp(-2Ax) - \dfrac{4(A+2)^2}{3A}\exp(-3Ax)\end{array}\right]}{\dfrac{x}{3}[5 + 3\exp(-2Ax) + \exp(-3Ax)]} \tag{4.44}$$

where $x = \lambda/kT$.

The behaviour of $\bar{\mu}$ versus $kT/|\lambda|$ calculated from Equation 4.44 for $A = 1\cdot0$, is shown in Fig. 4.9. Comparing this variation with that in the weak field case ($A = 1\cdot5$), we see that in general terms it follows a similar pattern for λ both positive and negative. The major differences are that in the strong field case, when λ is positive, $\overline{\mu}$ falls more slowly with temperature and tends towards a value of 1·23 B.M. at 0 K, whilst for λ negative the values of $\overline{\mu}$ at the higher temperatures are lower than in the weak field case.

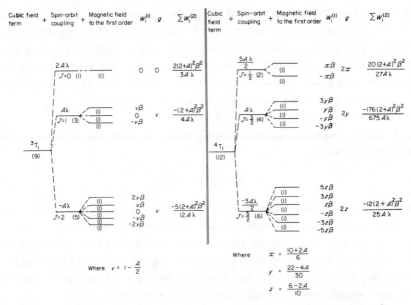

Figure 4.17. The splitting diagrams for the medium and strong field $^3T_1$ and $^4T_1$ terms under the action of spin–orbit coupling and a magnetic field.

$^4T_1$ terms

Spin–orbit coupling splits this term into three components at energies $-(3A\lambda)/2$, $A\lambda$, and $(5A\lambda)/2$, see Fig. 4.17. When $A = 1·5$ this splitting pattern corresponds to that of the weak field case given in Fig. 4.12. The results of the relevant perturbation calculation are summarized in Fig. 4.17 and in the equation:

$$\overline{\mu}^2 = \frac{\left[\dfrac{7(3-A)^2 x}{5} + \dfrac{12(A+2)^2}{25A} + \left\{\dfrac{2(11-2A)^2 x}{45} + \dfrac{176(A+2)^2}{675A}\right\} x\exp\left(-\dfrac{5Ax}{2}\right) + \left\{\dfrac{(A+5)^2 x}{9} - \dfrac{20(A+2)^2}{27A}\right\}\exp\left(-4Ax\right)\right]}{\dfrac{x}{3}\left[3 + 2\exp\left(-\dfrac{5Ax}{2}\right) + \exp\left(-4Ax\right)\right]}.$$

$$(4.45)$$

The behaviour of $\overline{\mu}$ against $kT/|\lambda|$ calculated from Equation 4.45 for $A = 1·0$, is shown in Fig. 4.13. The main difference between $\overline{\mu}$ for the weak and strong crystal field cases, for both positive and negative values of λ, is essentially that of magnitude rather than the rate of change with $kT/|\lambda|$.

4.5. The calculation of the magnetic properties of transition metal ions with T ground terms from a consideration of $t_2^n e^m$ configurations

In this section the calculation of the magnetic properties that arise solely from the T ground terms will be approached by taking the $t_2^n e^m$ configurations as a starting point. This approach has led to the development of closed form expressions for the effects of spin-orbit coupling and for the first- and second-order Zeeman coefficients [3]. In the present context of cubic crystal field terms these closed form expressions allow a good deal of the tedious perturbation calculation to be avoided.

In the limit of a strong crystal field the orbital wave-functions for the T terms may be written as

$$\psi(T) = \psi[T(t_2^n e^m)], \qquad (4.46)$$

where n and m are the numbers of electrons in each of the t_2 and e orbital sets.

At the beginning of this chapter it was demonstrated that the e set of orbitals has no orbital angular momentum associated with it. This means that as far as the calculation of the effects of spin–orbit coupling and a magnetic field are concerned the *orbital part* of the wave-function for the e set may be omitted. The total wave-function, including spin, may now be written as:

$$\psi(T) = \psi[T(t_2^n)]\psi(S), \qquad (4.47)$$

where $\psi(S)$ represents the appropriate spin wave-functions for the *term*.

It has been shown by Griffith [3, 4] that, as far as orbital angular momentum is concerned, there is a good deal of similarity between the t_2-set and the free ion p-set of orbitals. Using the following correspondence between the t_2- and p-orbital sets,

t_2-orbitals $\lvert l, m_l \rangle$	p-orbitals $\lvert l, m_l \rangle$
$\lvert 2, -1 \rangle$	$\equiv \lvert 1, 1 \rangle$
$-\sqrt{\tfrac{1}{2}}(\lvert 2, 2 \rangle - \lvert 2, -2 \rangle) \equiv \lvert 1, 0 \rangle$	
$-\lvert 2, 1 \rangle$	$\equiv \lvert 1, -1 \rangle$

it is easy to show that the matrix elements for orbital angular momentum between members of the p-orbital set are opposite in sign but equal in magnitude to those between the corresponding members of the t_2 set. This is equivalent to saying that the matrix elements of $\hat{l}$ between the t_2 set of orbitals are equal to those of $-\hat{l}$ between the corresponding p-orbitals as defined above. In general using these arguments Griffith showed that the effects of spin–orbit coupling and magnetic field on any t_2^n configuration, for $n \leqslant 6$, can be obtained from the free ion p^n configuration by replacing $\hat{L}$ by $\gamma\hat{L}$, where $\gamma = -1$. In effect we can set up a Russell-Saunders type coupling scheme within the t_2 orbital set using the effective quantum number $l' = 1$ for the t_2 electrons. The splitting of the t_2^n configuration due to spin–orbit coupling can now be obtained by applying the usual rules of spin–orbit coupling to give total orbital angular momentum quantum numbers, J, just as we would in the case of a free ion. This method gives the correct J values but compared to the free ion p^n configuration the set of multiplets is inverted, i.e. $J = L' + S$ is now lowest for a t_2 shell less than half full, rather than $J = |L - S|$ for the free ion p shell which is less than half full.

The $t_2^n - p^n$ correspondence can be extended to give closed form expressions for the g values and second-order Zeeman coefficients of the various spin–orbit levels. The general equation for the g value of any component of the multiplet arising from the action of the spin–orbit coupling is

$$g = \gamma + (2 - \gamma) \left[\frac{J(J + 1) + S(S + 1) - L(L + 1)}{2J(J + 1)} \right], \qquad (4.48)$$

unless $J = 0$, when g must be zero since the magnetic field does not split this level. For a free ion set of orbitals $\gamma = 1$ and Equation 4.48 reduces to the familiar Landé splitting factor, but for the t_2 set of orbitals $\gamma = -1$ and $L = L'$ so that

$$g(t_2) = -1 + 3 \left[\frac{J(J + 1) + S(S + 1) - L'(L' + 1)}{2J(J + 1)} \right]. \qquad (4.49)$$

In addition to the g value given by Equation 4.49 there is the following closed form expression for the *total* second order Zeeman coefficient of a level specified by J, due to the levels both *above* and *below* it, if both are present:

$$\sum_i W_i^{(2)}(J) = \frac{\beta^2 (g - \gamma)(g - 2)(2J + 1)}{-3\gamma\lambda}, \qquad (4.50)$$

where $\gamma = -1$ for the t_2 orbitals and when $J = 0$, g in this expression *only* is $4 - \gamma$.

In previous sections dealing with $^3T_1$ and $^4T_1$ terms the magnetic properties were dependent on the crystal field strength which was expressed in terms of the parameter A. These terms can also arise from certain $t_2^n e^m$ configurations and closed form expressions for the g values and second-order Zeeman coefficients can be written for any crystal field strength by replacing γ by $-A$ in Equations 4.48 and 4.50, i.e.

$$g = 1 - \frac{A}{2} + (2 + A) \left[\frac{S(S + 1) - L'(L' + 1)}{2J(J + 1)} \right], \qquad (4.51)$$

except for $J = 0$ when $g = 0$, and

$$\sum_i W_i^{(2)}(J) = \frac{\beta^2 (g + A)(g - 2)(2J + 1)}{3A\lambda},$$ (4.52)

where $g = 4 + A$ has to be used when $J = 0$.

The usefulness of Equations 4.48 to 4.52 is best illustrated with a few examples, but the information for all the possible orbital triplet ground states is summarized in Figs. 4.2, 4.5, and 4.17.

Example 1

$^3T_1$ term which can arise from the t_2^2 configuration.

Step 1. Determination of L' and S.

$$m_l = 1 \quad 0 \quad -1$$

↑	↑	

Since for t_2 orbitals $l' = 1$, the possible m_l values for the electrons are $1, 0, -1$. The L' value for the $^3T_1$ term from the t_2^2 configuration is equal to the maximum value of Σm_l, i.e. $L' = 1$. The value of S corresponding to this L' is $S = 1$.

Step 2. Determination of the splitting due to spin–orbit coupling. Remembering that $J = L + S, L + S - 1, L + S - 2 \ldots |L - S|$ we have for the t_2^2 configuration $J = 2, 1, 0$ and spin–orbit coupling results in a splitting into three components, see Fig. 4.17. The energy difference between the level specified by J and that by $(J + 1)$ is $-A(J + 1)\lambda$ and the energies relative to the un-split level can be found from this and the centre of gravity rule. If we let $E(J)$ be the energy of the level specified by J and $n(J)$ its total multiplicity (i.e. $2J + 1$) then $\Sigma_J n(J) . E(J) = 0$. Applying this to the current problem we have:

$$\left.\begin{array}{l} (5)(E(2)) + (3)(E(1)) + E(0) = 0 \\ E(0) - E(1) = A\lambda \\ E(1) - E(2) = 2A\lambda \end{array}\right\}.$$ (4.53)

The solutions to these equations are

$$E(2) = -A\lambda, E(1) = A\lambda, \text{ and } E(0) = 2A\lambda.$$

Step 3. The g values and second-order Zeeman coefficients for each spin–orbit level are now calculated using Equations 4.51 and 4.52.

(i) For the level $J = 0, g = 0$ and $\sum_i W_i^{(2)}(0) = \frac{\beta^2}{\lambda} \frac{2}{3} \frac{(A + 2)^2}{A}$

(ii) For $J = 1, g = 1 - \dfrac{A}{2}$ and $\sum W_i^{(2)}(1) = -\dfrac{\beta^2}{\lambda}\dfrac{(2+A)^2}{4A}$

(iii) For $J = 2, g = 1 - \dfrac{A}{2}$ and $\sum W_i^{(2)}(2) = -\dfrac{\beta^2}{\lambda}\dfrac{5}{12}\dfrac{(2+A)^2}{A}$

These g and $\sum W_i^{(2)}(J)$ values coupled with the energies calculated in step 2 can now be used to calculate the magnetic susceptibility in the normal way using Van Vleck's equation.

Example 2

The $^4T_1$ term which can arise from the $e^2 t_2^1$ configuration.

Step 1. As mentioned previously we may ignore the orbital wave-functions for the e set since they contribute nothing to our final result. The orbital angular momentum is contributed solely by the t_2^1 part of the configuration and thus we can work out L' in the same way as in the previous example, i.e.

$$m_l = 1 \quad 0 \quad -1$$

↑		

for which $L' = \max \sum m_l = 1$. However, as far as the spin angular momentum is concerned we must take account of all the unpaired electrons, which in this case gives $S = \frac{3}{2}$.

Step 2. The effects of spin–orbit coupling are now calculated using $L' = 1$ and $S = \frac{3}{2}$. This gives $J = \frac{5}{2}, \frac{3}{2}, \frac{1}{2}$, at energies $-3A\lambda/2$, $A\lambda$, and $5A\lambda/2$ relative to the unperturbed level.

Step 3. The g values and second-order Zeeman coefficients for each spin–orbit level are calculated using Equations 4.51 and 4.52.

(i) For $J = \frac{1}{2}, g = \dfrac{10 + 2A}{3}$, and $\sum W_i^{(2)}(\frac{1}{2}) = \dfrac{20}{27}\dfrac{(2+A)^2}{A}\dfrac{\beta^2}{\lambda}$.

(ii) For $J = \frac{3}{2}, g = \dfrac{22 - 4A}{15}$, and $\sum W_i^{(2)}(\frac{3}{2}) = -\dfrac{176}{675}\dfrac{(2+A)^2}{A}\dfrac{\beta^2}{\lambda}$.

(iii) For $J = \frac{5}{2}, g = \dfrac{6 - 2A}{5}$, and $\sum W_i^{(2)}(\frac{5}{2}) = -\dfrac{12}{25}\dfrac{(2+A)^2}{A}\dfrac{\beta^2}{\lambda}$.

Example 3

The $^3A_2$ term which arises from the t_2^3 configuration.

Step 1. The determination of L' and S.

$$m_l = 1 \quad 0 \quad -1$$

↑	↑	↑

for which $L' = \max \sum m_l = 0$ and $S = \frac{3}{2}$.

Step 2. The calculation of the effects of spin–orbit coupling. With $L' = 0$ and $S = \frac{3}{2}$ the only possible value of J is $\frac{3}{2}$, which means that spin–orbit coupling does not remove any of the degeneracy of this configuration. This is precisely the result we obtained from our previous consideration.

Step 3. The calculation of the g values and second order Zeeman coefficients. In this case Equations 4.49 and 4.50 are used, since the variation in crystal field strength does not modify the wave-functions of the term we are considering. Thus we obtain $g = 2$ and $\Sigma W_i^{(2)}(\frac{3}{2}) = 0$, which is precisely the result we would have obtained from our more extensive perturbation calculation if we had ignored the 'mixing in', under spin–orbit coupling, of the excited orbitally degenerate $^3T_2$ term.

These three calculations demonstrate the usefulness and also the limitations of the $t_2^n - p^n$ equivalence approach. At this stage some readers may question the need for the detailed presentation of the more lengthy perturbation approach to the calculation of the magnetic susceptibilities, when so many of the required results for octahedral or tetrahedral complexes can be obtained from the $t_2^n - p^n$ equivalence. The reasons for our approach are that the above considerations are restricted to the contributions from the *ground term* in *perfect octahedral* and *tetrahedral* stereochemistries. In anticipation of the requirements of the following chapter, when non-cubic stereochemistries are considered, where the $t_2^n - p^n$ equivalence cannot be applied, the perturbation approach has been illustrated with the relatively simple examples of the $^2T_2$ and $^3A_2$ terms.

4.6. The magnetic properties of spin-paired transition metal complexes in cubic crystal fields

When the crystal field is sufficiently strong to force spin-pairing of the d-electrons, the ground states are expressed as $t_2^n e^m$ configurations which give rise to various terms when the action of interelectronic repulsions is considered. The ground terms which arise for various $t_2^n e^m$ configurations in cubic crystal fields are given in Table 4.4.

The magnetic properties of the terms which arise in spin-paired complexes are exactly those given in the previous sections, providing the appropriate values of the spin–orbit coupling constant, λ, and where appropriate the crystal field parameter, A, are used. The spin–orbit coupling constant for the ground term is given by:

$$\lambda = \pm \zeta_{nd}/2S, \tag{4.54}$$

where ζ_{nd} is the single electron spin–orbit coupling constant for the set of d-electrons with principal quantum number n.

S is the total spin angular momentum quantum number for this term.

The minus sign applies when the t_2 shell is more than half full.

Table 4.4. *Ground terms which arise from various $t_2^n e^m$ configurations.*

No. of d-electrons	Stereo-chemistry	Strong field ground configuration	Strong field ground term
1	Octahedral	t_{2g}^1	$^2T_{2g}$
	Tetrahedral	e^1	^2E
2	Octahedral	t_{2g}^2	$^3T_{1g}$
	Tetrahedral	e^2	$^3A_2$
3	Octahedral	t_{2g}^3	$^4A_{2g}$
	Tetrahedral	$e^2t_2^1$	$^4T_1$
4	Octahedral	$t_{2g}^3e_g^1$	^5E_g
	Octahedral	t_{2g}^4	$^3T_{1g}$
	Tetrahedral	$e^2t_2^2$	$^5T_2$
5	Octahedral	$t_{2g}^3e_g^2$	$^6A_{1g}$
	Octahedral	t_{2g}^5	$^2T_{2g}$
	Tetrahedral	$e^2t_2^3$	$^6A_1$
6	Octahedral	$t_{2g}^4e_g^2$	$^5T_{2g}$
	Octahedral	t_{2g}^6	$^1A_{1g}$
	Tetrahedral	$e^3t_2^3$	^5E
7	Octahedral	$t_{2g}^5e_g^2$	$^4T_{1g}$
	Octahedral	$t_{2g}^6e_g^1$	^2E_g
	Tetrahedral	$e^4t_2^3$	$^4A_2$
8	Octahedral	$t_{2g}^6e_g^2$	$^3A_{2g}$
	Tetrahedral	$e^4t_2^4$	$^3T_1$
9	Octahedral	$t_{2g}^6e_g^3$	^2E_g
	Tetrahedral	$e^4t_2^5$	$^2T_2$

Concluding remarks

In this chapter we have shown how the magnetic susceptibilities of transition metal complexes may be calculated. The variation of the magnetic susceptibilities with temperature of a number of transition metal ions in perfectly octahedral and tetrahedral stereochemistries has been discussed in general terms but no comparison with actual compounds has been attempted. This comparison is postponed until Chapter 6 for the following reasons. So far we have considered the transition metal complexes purely from the crystal field approach in that only the metal d-orbitals have been considered. It is now well established that this approximation is inadequate and some form of allowance for covalency is usually necessary. Also for a number of reasons, amongst which are the Jahn-Teller effect and crystal packing considerations, complexes which have perfect tetrahedral or octahedral stereochemistries are extremely rare. However, an understanding of the magnetic behaviour of these systems provides an essential springboard for the more complicated cases of non-cubic stereochemistries. Some of the consequences of deviation from perfect cubic symmetry and of covalent bonding are discussed in more detail in the next chapter.

REFERENCES

1. B.N. Figgis (1958). *Nature*, **182**, 1568, and references therein.
2. B.N. Figgis (1965). *J. Chem. Soc.*, 4887; B.N. Figgis, J. Lewis, F.E. Mabbs and G.A. Webb (1966). *J. Chem. Soc. (A)*, 1411.
3. J.S. Griffith (1964). *The Theory of Transition Metal Ions*, Cambridge University Press, page 239.
4. M. Gerloch and J.R. Miller (1968). *Progr. Inorg. Chem.*, **10**, 1.

5

The magnetic properties of transition metal ions in axially symmetric crystal fields

In the previous chapter the magnetic properties of cubic field $^2T_2$ terms were calculated in detail and the results for the other transition metal ion ground terms quoted. However, the number of transition metal complexes which have regular octahedral or tetrahedral stereochemistry is very small indeed. The consequence of this is that the previous model is very restricted in its applicability and it must be extended so that it approaches more closely the situation in actual complexes. The method of dealing with this problem will be illustrated in detail for the $^2T_2$ ground term when it is perturbed by a tetragonal crystal field component and spin–orbit coupling, and when allowance is made for possible reductions in orbital angular momentum.

The effect of the departure from cubic symmetry is to make the system anisotropic, e.g. in an axially symmetric system $\chi_z \neq \chi_x = \chi_y$. In such systems it is now essential to measure the principal magnetic susceptibilities in order to make a meaningful comparison between the experimental data and that calculated assuming a particular model for the complex. Before proceeding with the details of the calculation for the $^2T_2$ case a brief outline of the effects of covalency on orbital angular momentum will be given.

5.1. The reduction of orbital angular momentum

The concept of a reduction in the orbital angular momentum of the unpaired electrons in a complex stems from the classical work of Griffiths et al. and Stevens [1] on the electron spin resonance of $IrCl_6^{2-}$. They found that the experimental g value was significantly lower than that calculated from the model which only considered the metal d-orbitals. The need to include ligand orbitals

107

was demonstrated by the observation of chlorine superhyperfine-splittings in the electron spin resonance spectrum of this complex, which shows that the unpaired electron spends some of its time close to the nucleus of the chloride ligands. It was shown that the formation of molecular orbitals involving the metal d- and ligand p-orbitals leads to a reduction in the orbital angular momentum of the unpaired electron compared with the simple case involving only the pure metal d-orbitals. Stevens also showed that the same effect could be obtained by using the pure metal d-orbitals and a modified expression for the magnetic moment operator *viz.*:

$$\hat{\mu}H = \beta H(k\hat{L} + 2\hat{S}). \tag{5.1}$$

The quantity **k** may be regarded as a parameter which allows for the reduction in the orbital angular momentum. It is sometimes referred to as an electron delocalization parameter, but the term 'orbital reduction factor' is more appropriate since we will see later that ligand contributions may not be the only cause of a reduction in orbital angular momentum.

There are two excellent reviews [2] which deal with the quantitative aspects of **k** and hence only a brief and simplified treatment of the relationship between **k** and the reduction in orbital angular momentum will be given here.

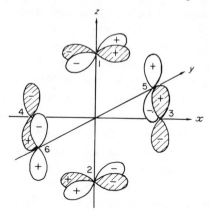

Figure 5.1. A three-dimensional view of ligand atomic orbitals which overlap with d_{xz} and d_{yz} metal orbitals in O_h. (Reproduced with permission from Gerloch and Miller [2].)

If in an octahedral complex we consider the formation of molecular orbitals for the d_{yz} and d_{xz} orbitals, (see Fig. 5.1) of the t_{2g} set we have:

$$|xz\rangle = N\left[d_{xz} + \frac{\alpha}{2}(p_x^{(1)} - p_x^{(2)} + p_z^{(3)} - p_z^{(4)})\right]$$

$$|yz\rangle = N\left[d_{yz} + \frac{\alpha}{2}(p_y^{(1)} - p_y^{(2)} + p_z^{(5)} - p_z^{(6)})\right]$$

where α is a mixing coefficient.

The orbital angular momentum about the z-axis associated with this pair of orbitals is:

$$\langle xz|\hat{l}_z|yz\rangle = N^2 \left[\left\langle d_{xz} + \frac{\alpha}{2}(p_x^{(1)} - p_x^{(2)} + p_z^{(3)} - p_z^{(4)}) \left| \hat{l}_z \right| d_{yz} \right. \right.$$
$$\left. \left. + \frac{\alpha}{2}(p_y^{(1)} - p_y^{(2)} + p_z^{(5)} - p_z^{(6)}) \right\rangle \right]$$

where $N^2 = (1 + \alpha^2 + 4\alpha S_{d,p})^{-1}$ if ligand-ligand overlap is ignored; and $S_{d,p}$ is the overlap integral between a metal d-orbital and a single ligand p-orbital.

If we again ignore ligand-ligand terms we have

$$\langle xz|\hat{l}_z|yz\rangle = N^2 \left[\begin{array}{l} \langle d_{xz}|\hat{l}_z|d_{yz}\rangle \\[2mm] + \dfrac{\alpha}{2}\langle(p_x^{(1)} - p_x^{(2)} + p_z^{(3)} - p_z^{(4)})|\hat{l}_z|d_{yz}\rangle \\[2mm] + \dfrac{\alpha}{2}\langle d_{xz}|\hat{l}_z|(p_y^{(1)} - p_y^{(2)} + p_z^{(5)} - p_z^{(6)})\rangle \end{array} \right. \begin{array}{l} 1 \\[2mm] 2 \\[2mm] 3 \end{array} \quad (5.2)$$

Evaluating (5.2) term by term:

Term 1: from Table 4.1 this gives $-i$

Term 2:

$$\langle(p_x^{(1)} - p_x^{(2)} + p_z^{(3)} - p_z^{(4)})|\hat{l}_z|d_{yz}\rangle$$
$$= \langle(p_x^{(1)} - p_x^{(2)} + p_z^{(3)} - p_z^{(4)}|-i\,d_{xz}\rangle = -4iS_{d,p}$$

Term 3: this term is evaluated using the Hermitian† properties of $\hat{l}_z$. Thus

$$\langle d_{xz}|\hat{l}_z|(p_y^{(1)} - p_y^{(2)} + p_z^{(5)} - p_z^{(6)}\rangle$$
$$= \langle(p_y^{(1)} - p_y^{(2)} + p_z^{(5)} - p_z^{(6)})|\hat{l}_z^*|d_{xz}^*\rangle$$
$$= -\langle(p_y^{(1)} - p_y^{(2)} + p_z^{(5)} - p_z^{(6)})|\hat{l}_z|d_{xz}\rangle = -4iS_{d,p}.$$

since $|d_{xz}^*\rangle = |d_{xz}\rangle$ (because this is a real orbital) and

$$\hat{l}_z^* = -\frac{ih}{2\pi}\left(x\frac{\partial}{\partial y} - y\frac{\partial}{\partial x}\right) = -\hat{l}_z.$$

The summation of the individual contributions gives:

$$\langle xz|\hat{l}_z|yz\rangle = -iN^2(1 + 4\alpha S_{d,p}). \qquad (5.3)$$

If, instead of using the molecular orbital approach, we use the modified form of the orbital angular momentum operator we get

$$\langle d_{xz}|k\hat{l}_z|d_{yz}\rangle = -ki \qquad (5.4)$$

---

† An operator is said to be Hermitian if $\int \phi^* \mathcal{H} \psi \, d\tau = \int \psi \mathcal{H}^* \phi^* d\tau$. For further discussion see H. Eyring, J. Walter, and E.B. Kimball, *Quantum Chemistry*, page 27.

Equating (5.3) and (5.4) we find

$$k = N^2(1 + 4\alpha S_{d,p}) = 1 - \alpha^2 N^2. \tag{5.5}$$

Since α^2 and N^2 are positive quantities Equation 5.5 shows that **k** is less than unity. In other words the formation of molecular orbitals in this particular example leads to a lowering of the orbital angular momentum compared with that for the pure metal ion d-orbitals. Indeed, more extensive considerations show that for molecular orbital formation **k** is always less than or equal to unity.

In non-centrosymmetric transition metal complexes, for example those with T_d symmetry, as well as a reduction in the orbital angular momentum being caused by ligand mixing, metal d–p orbital mixing also has a similar effect. If the functions

$$|xz\rangle = N\,d_{xz} + np_y$$

and

$$|yz\rangle = N\,d_{yz} + np_x$$

are taken as a basis set, it is found that $k = (N^2 - n^2)$. This makes $k < 1$ since $0 < N^2 < 1$ and $0 < n^2 < 1$.

5.2. The magnetic properties of T terms in axially symmetric crystal fields

5.2.1. THE MAGNETIC PROPERTIES OF $^2T_2$ TERMS IN AXIALLY SYMMETRIC CRYSTAL FIELDS [3]

The method most frequently used to calculate the magnetic susceptibilities of the $^2T_2$ term under the action of an axially symmetric crystal field and spin–orbit coupling is that of perturbation theory. Using this method the magnetic susceptibilities of the $^2T_2$ term in a tetragonal crystal field will now be given in detail. Although all the perturbations in the system could be applied simultaneously it is far more convenient to treat the effects of the tetragonal crystal field and spin–orbit coupling as simultaneous perturbations. The effect of a magnetic field, which causes changes in energy which are considerably smaller than the previous perturbations, is then treated as a further perturbation.

The first stage in the calculation is to find the energy levels and wave-functions which result from the simultaneous perturbation of the $^2T_2$ term by spin–orbit coupling and the tetragonal crystal field component, represented here by the potential V_{tetrag}. Thus we need to solve the energy equation

$$(\mathcal{H}_0 + \mathcal{H}^{(1)})\psi = E\psi, \tag{5.6}$$

where $\mathcal{H}_0$ is the Hamiltonian for the system as far as the cubic crystal field is concerned, and $\mathcal{H}^{(1)} = \lambda \hat{L} \cdot \hat{S} + V_{\text{tetrag.}}$, which represents the part of the Hamiltonian for the perturbation by spin–orbit coupling and the tetragonal crystal field component.

The problem has been solved for $\mathscr{H}_0$ and the wave-functions for the $^2T_2$ ground term have been given in Section 4.3.1. Since the perturbation $\mathscr{H}^{(1)}$ is acting on a degenerate term we need to construct the secular determinant involving the $^2T_2$ wave-functions analogous to Equation 4.10. The matrix elements involved are of the type $\langle\psi_i|\lambda\hat{L}.\hat{S} + V_{\text{tetrag.}}|\psi_j\rangle$ and one example of the evaluation of such matrix elements will now be given, the evaluation of the remainder being left as an exercise for the reader.

Example

$$\langle\psi_2|\lambda\hat{L}.\hat{S} + V_{\text{tetrag}}|\psi_2\rangle = \langle\psi_2|\lambda\hat{L}.\hat{S}|\psi_2\rangle + \langle\psi_2|V_{\text{tetrag}}|\psi_2\rangle.$$

The first of these matrix elements was evaluated in Section 4.3.1 and the second since it involves components of the orbital doublet will be at an energy Δ above the orbital singlet (see Chapter 2)

$$\langle\psi_2|\lambda\hat{L}.\hat{S} + V_{\text{tetrag.}}|\psi_2\rangle = -\frac{\lambda}{2} + \Delta.$$

Evaluation of all the remaining matrix elements leads to the following secular determinant:

	$|\psi_2\rangle$	$|\psi_3\rangle$	$|\psi_4\rangle$	$|\psi_5\rangle$	$|\psi_1\rangle$	$|\psi_6\rangle$
$\langle\psi_2|$	$-\dfrac{\lambda}{2}+\Delta-E$	0	0	0	0	0
$\langle\psi_3|$	0	$-\dfrac{\lambda}{2}+\Delta-E$	0	0	0	0
$\langle\psi_4|$	0	0	$\dfrac{\lambda}{2}+\Delta-E$	$-\lambda/\sqrt{2}$	0	0
$\langle\psi_5|$	0	0	$-\lambda/\sqrt{2}$	$0-E$	0	0
$\langle\psi_1|$	0	0	0	0	$\dfrac{\lambda}{2}+\Delta-E$	$\lambda/\sqrt{2}$
$\langle\psi_6|$	0	0	0	0	$\lambda/\sqrt{2}$	$0-E$

$= 0.$

(5.7)

The energies and zero order wave-functions may be obtained from (5.7) by proceeding as in Section 4.3.1. Thus we have the following four sub-determinants to solve:

$$
\begin{array}{c}
|\psi_2\rangle \\
\langle\psi_2| \quad \left| -\dfrac{\lambda}{2}+\Delta-E \right| \quad = 0,
\end{array}
$$

i.e.

$$E_1 = \Delta - (\lambda/2). \tag{5.8}$$

$$
\begin{array}{c}
\quad |\psi_3\rangle \\
\hline
\langle\psi_3| \quad \left| -\dfrac{\lambda}{2} + \Delta - E \right| \quad = 0,
\end{array}
$$

i.e.

$$E_2 = \Delta - (\lambda/2). \tag{5.9}$$

$$
\begin{array}{c|cc}
 & |\psi_4\rangle & |\psi_5\rangle \\
\hline
\langle\psi_4| & \dfrac{\lambda}{2} + \Delta - E & -\lambda/\sqrt{2} \\
\langle\psi_5| & -\lambda/\sqrt{2} & 0 - E
\end{array}
\quad = 0,
$$

i.e.

$$E_3 = \frac{1}{2}\left[\left(\frac{\lambda}{2} + \Delta\right) + \left(\Delta^2 + \lambda\Delta + \frac{9\lambda^2}{4}\right)^{\frac{1}{2}}\right], \tag{5.10}$$

$$E_4 = \frac{1}{2}\left[\left(\frac{\lambda}{2} + \Delta\right) - \left(\Delta^2 + \lambda\Delta + \frac{9\lambda^2}{4}\right)^{\frac{1}{2}}\right], \tag{5.11}$$

$$
\begin{array}{c|cc}
 & |\psi_1\rangle & |\psi_6\rangle \\
\hline
\langle\psi_1| & \dfrac{\lambda}{2} + \Delta - E & \lambda/\sqrt{2} \\
\langle\psi_6| & \lambda/\sqrt{2} & 0 - E
\end{array}
\quad = 0.
$$

i.e.

$$E_5 = \frac{1}{2}\left[\left(\frac{\lambda}{2} + \Delta\right) + \left(\Delta^2 + \lambda\Delta + \frac{9\lambda^2}{4}\right)^{\frac{1}{2}}\right], \tag{5.12}$$

$$E_6 = \frac{1}{2}\left[\left(\frac{\lambda}{2} + \Delta\right) - \left(\Delta^2 + \lambda\Delta + \frac{9\lambda^2}{4}\right)^{\frac{1}{2}}\right], \tag{5.13}$$

As in Section 4.3.1 these energies are substituted into the sub-determinants from which they were obtained in order to get the required zero order wave-functions. Whilst doing this it is convenient to adopt the notation of Figgis [3] and put $v = \Delta/\lambda$. The energies and corresponding zero order wave-functions are summarized in Table 5.1

The results so far show that the combined effect of perturbation by spin–orbit coupling and a tetragonal crystal field component is to split the originally six-fold degenerate $^2T_2$ term into a set of three doubly degenerate

Table 5.1. *Energies and wave-functions for the* $^2T_2$ *term after perturbation by spin-orbit coupling and a tetragonal crystal field component.*

Energy	Zero-order wave-function
$E_1 = \lambda(v - \frac{1}{2})$	$\phi_1 = \psi_2$
$E_2 = \lambda(v - \frac{1}{2})$	$\phi_2 = \psi_3$
$E_3 = \dfrac{\lambda}{2}[(\frac{1}{2} + v) + (v^2 + v + \frac{9}{4})^{\frac{1}{2}}]$	$\phi_3 = (1 + \alpha^2)^{-\frac{1}{2}}[\psi_4 + \alpha\psi_5]$
$E_4 = \dfrac{\lambda}{2}[(\frac{1}{2} + v) - (v^2 + v + \frac{9}{4})^{\frac{1}{2}}]$	$\phi_4 = (1 + b^2)^{-\frac{1}{2}}[\psi_4 + b\psi_5]$
$E_5 = \dfrac{\lambda}{2}[(\frac{1}{2} + v) + (v^2 + v + \frac{9}{4})^{\frac{1}{2}}]$	$\phi_5 = (1 + \alpha^2)^{-\frac{1}{2}}[\psi_1 - \alpha\psi_6]$
$E_6 = \dfrac{\lambda}{2}[(\frac{1}{2} + v) - (v^2 + v + \frac{9}{4})^{\frac{1}{2}}]$	$\phi_6 = (1 + b^2)^{-\frac{1}{2}}[\psi_1 - b\psi_6]$

Where $\alpha = [(\frac{1}{2} + v) - (v^2 + v + \frac{9}{4})^{\frac{1}{2}}]/\sqrt{2}$, and $b = [(\frac{1}{2} + v) + (v^2 + v + \frac{9}{4})^{\frac{1}{2}}]/\sqrt{2}$.

levels. Now that the energies and wave-functions resulting from the perturbation by the low symmetry crystal field and by spin–orbit coupling have been obtained, we are in a position to consider the effect of the magnetic field perturbation. Because the system is anisotropic the effect of applying a magnetic field in the z or parallel direction of the molecule will be different from that obtained if it is applied in the perpendicular direction. The problem, therefore, is one of finding the changes in the energies of the three doublets when the magnetic field is applied firstly in the parallel and then in the perpendicular direction.

At this stage in the calculation it is convenient to include the possibility of a loss in orbital angular momentum by including the orbital reduction parameter **k**. Since we have so far assumed that the effects of spin–orbit coupling are isotropic, we will do the same for the reduction of the orbital angular momentum. Although this may not often be a good approximation it is useful when illustrating the method of calculation since it only requires a single parameter. The terms in the Hamiltonian which express the magnetic field perturbation are written as $(k\hat{L}_z + 2\hat{S}_z)\beta H_z$ and $(k\hat{L}_x + 2\hat{S}_x)\beta H_x$ for the parallel and perpendicular directions respectively (since the system is axially symmetric we need only calculate the result in the x-direction, that for the y-direction being identical).

The calculation for each of the two principal directions takes the same form as for the isotropic case in Chapter 4. Again any interaction in the magnetic field between levels arising from the original ^2E term and those from the split $^2T_2$

term will be neglected. Since we have a set of three doubly degenerate levels the first-order Zeeman coefficients are found for each individual level by constructing the appropriate secular determinants. The second-order Zeeman coefficients are then found by evaluating the matrix elements of the magnetic moment operator between levels which were not degenerate with each other in the absence of the magnetic field. The magnetic susceptibilities are then calculated by substituting these coefficients in Van Vleck's susceptibility equation.

The calculation of $\chi_{\parallel}$

Only a few of the Zeeman coefficients will be calculated in detail in order to illustrate the method of calculation, the remainder are summarized in Table 5.2.

Example

The calculation of the first-order Zeeman coefficients for the doublet at energy E_1. Since we are dealing with a degenerate level, the secular determinant involving the matrix elements between ϕ_1 and ϕ_2 under the action of the operator $(k\hat{L}_z + 2\hat{S}_z)\beta H_z$ has to be constructed and solved:

$$
\begin{array}{c|cc}
 & |\phi_1\rangle & |\phi_2\rangle \\
\hline
\langle\phi_1| & (k-1)\beta H_z - E & 0 \\
\langle\phi_2| & 0 & -(k-1)\beta H_z - E
\end{array} = 0. \quad (5.14)
$$

The solutions to this determinant are

$$E = \pm(k-1)\beta H_z$$

i.e. the first order Zeeman coefficients are

$$W_1^{(1)}(z) = \pm(k-1)\beta. \quad (5.15)*$$

Example

The calculation of the second-order Zeeman coefficients. These coefficients are given by evaluating the following expressions, the results being summarized in Table 5.2

(i) the total second-order Zeeman coefficient for the doublet at E_1 is given by:*

$$\sum_i \sum_j \frac{\langle\phi_i|k\hat{L}_z + 2\hat{S}_z)\beta H_z|\phi_j\rangle\langle\phi_j|k\hat{L}_z + 2\hat{S}_z)\beta H_z|\phi_i\rangle}{E_i - E_j}. \quad (5.16)$$

where i has the values 1 and 2 and j has the values 3, 4, 5, and 6.

---

* The first- and second-order Zeeman coefficients $W_i^{(1)}$ and $W_i^{(2)}$ respectively for any level will be specified by one or other of the values of i which were used to label the energy level before the action of the magnetic field.

(ii) at energy E_3 the appropriate expression is given by (5.16) when:
i has the values 3 and 5, and
j has the values 1, 2, 4, and 6.
(iii) similarly at energy E_4, Equation 5.16 is appropriate when:
i has the values 4 and 6, and
j has the values 1, 2, 3, and 5.

Table 5.2. *First- and second-order Zeeman coefficients*

Energy E_i	$W_i^{(1)}(z)/\beta$	$W_i^{(2)}(z)/\beta$	$W_i^{(1)}(x)/\beta$	$W_i^{(2)}(x)/\beta$
$E_1 = E_2$	$\pm(k-1)$	0	0	$\dfrac{4}{\lambda}\left[\dfrac{\left(1-\dfrac{k\alpha}{\sqrt{2}}\right)^2}{(1+\alpha^2)(v-\frac{3}{2}-Z)} + \dfrac{\left(1-\dfrac{kb}{\sqrt{2}}\right)^2}{(1+b^2)(v-\frac{3}{2}+Z)}\right]$
$E_3 = E_5$	$\dfrac{\pm(k+1-\alpha^2)}{(1+\alpha^2)}$	$\dfrac{2(k+1-\alpha b)^2}{\lambda(1+\alpha^2)(1+b^2)Z}$	$\dfrac{\pm(\sqrt{2}k\alpha-\alpha^2)}{(1+\alpha^2)}$	$\dfrac{4}{\lambda}\left[\dfrac{\left(\dfrac{k\alpha}{\sqrt{2}}+\dfrac{kb}{\sqrt{2}}-\alpha b\right)^2}{2Z(1+\alpha^2)(1+b^2)} - \dfrac{\left(1-\dfrac{k\alpha}{\sqrt{2}}\right)^2}{(1+\alpha^2)(v-\frac{3}{2}-Z)}\right]$
$E_4 = E_6$	$\dfrac{\pm(k+1-b^2)}{(1+b^2)}$	$\dfrac{-2(k+1-\alpha b)^2}{\lambda(1+\alpha^2)(1+b^2)Z}$	$\dfrac{\pm(\sqrt{2}kb-b^2)}{(1+b^2)}$	$-\dfrac{4}{\lambda}\left[\dfrac{\left(\dfrac{k\alpha}{\sqrt{2}}+\dfrac{kb}{\sqrt{2}}-\alpha b\right)^2}{2Z(1+\alpha^2)(1+b^2)} + \dfrac{\left(1-\dfrac{kb}{\sqrt{2}}\right)^2}{(1+b^2)(v-\frac{3}{2}+Z)}\right]$

Where $Z = (v^2 + v + \frac{9}{4})^{\frac{1}{2}}$.

All the coefficients required to calculate $\chi_\parallel$ are known and substitution in Van Vleck's equation gives:

$$\chi_\parallel = \frac{N\beta^2}{3kT}3\left[\frac{\begin{array}{l}2\{W_1^{(1)}(z)\}^2\exp(-E_1/kT)+(2\{W_3^{(1)}(z)\}^2-2kTW_3^{(2)}(z))\\ x\exp(-E_3/kT)+(2\{W_4^{(1)}(z)\}^2-2kTW_4^{(2)}(z))\exp(-E_4/kT)\end{array}}{2\exp(-E_1/kT)+2\exp(-E_3/kT)+2\exp(-E_4/kT)}\right].$$

$$(5.17)$$

Calculation of $\chi_\perp$

As in the calculation for $\chi_\parallel$ only a few examples will be calculated in detail, all the required results being summarized in Table 5.2.

Example

The first-order Zeeman coefficients for the doublet at energy E_1.
    The appropriate secular determinant is:

$$
\begin{array}{c|cc}
 & |\phi_1\rangle & |\phi_2\rangle \\
\hline
\langle\phi_1| & 0-E & 0 \\
\langle\phi_2| & 0 & 0-E
\end{array} = 0, \qquad (5.18)
$$

i.e. $W_1^{(1)}(x) = 0$ which tells us that a magnetic field in this direction does not remove the degeneracy of this level.

Example

The first-order Zeeman coefficients for the doublet at energy E_3.
    The secular determinant in this case is:

$$
\begin{array}{c|cc}
 & |\phi_3\rangle & |\phi_5\rangle \\
\hline
\langle\phi_3| & 0-E & \dfrac{(\sqrt{2}k\alpha - \alpha^2)\beta H_x}{(1+\alpha^2)} \\[3mm]
\langle\phi_5| & \dfrac{(\sqrt{2}k\alpha - \alpha^2)\beta H_x}{(1+\alpha^2)} & 0-E
\end{array} = 0. \qquad (5.19)
$$

The solutions to this determinant are:

$$
E = \pm \frac{(\sqrt{2}k\alpha - \alpha^2)\beta H_x}{(1+\alpha^2)},
$$

and hence

$$
W_3^{(1)}(x) = \pm \frac{(\sqrt{2}k\alpha - \alpha^2)\beta}{(1+\alpha^2)}. \qquad (5.20)
$$

An interesting feature of this case is that in contrast to the parallel direction the determinant contains only non-zero *off diagonal* matrix elements. This means that not only does the magnetic field remove the two-fold degeneracy of this level, but it also mixes the functions $|\phi_3\rangle$ and $|\phi_5\rangle$. No such mixing occurs if there are only non-zero *diagonal* matrix elements.
    The second-order Zeeman coefficients are calculated using Equation 5.16 with the magnetic moment operator replaced by $(k\hat{L}_x + 2\hat{S}_x)\beta H_x$. The results

for the second-order Zeeman coefficients are summarized in Table 5.2. The substitution of the Zeeman coefficients into Van Vleck's equation gives:

$$
\chi_\perp = \frac{N\beta^2}{3kT} 3 \left[\frac{
\begin{array}{l}
-2kTW_1^{(2)}(x)\exp(-E_1/kT)+(2\{W_3^{(1)}(x)\}^2-2kTW_3^{(2)}(x)) \\
x\exp(-E_3/kT)+(2\{W_4^{(1)}(x)\}^2-2kTW_4^{(2)}(x))\exp(-E_4/kT)
\end{array}
}{2\exp(-E_1/kT)+2\exp(-E_3/kT)+2\exp(-E_4/kT)} \right].
$$

(5.21)

The average susceptibilities may readily be calculated using the relationship $\bar{\chi} = \frac{1}{3}(\chi_\parallel + 2\chi_\perp)$. However, it is more convenient to talk in terms of μ and the variation of this quantity with the parameters v, kT/λ, and $\mathbf{k}$ are given in Figs. 5.2–5.5.

The results obtained for the tetragonal distortions are applicable to a trigonally distorted octahedral situation provided we restrict our attention to the $^2T_2$ ground term, since it also will split into an orbital doublet and singlet. Indeed, although the origins of the splitting may be different all the results for T ground terms given in this chapter are equally applicable to tetragonally and trigonally distorted octahedra.

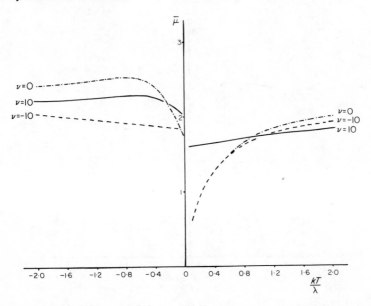

Figure 5.2. Diagrams of $\bar{\mu}$ vs. kT/λ for the $^2T_2$ term for various values of v, with $\mathbf{k} = 1\cdot0$.

(a) *The variation of $\bar{\mu}$ with v* (illustrated in Fig. 5.2). When λ is positive the values of $\bar{\mu}$ for $v = \pm10$ at the higher values of kT/λ are only slightly lower than for $v = 0$, and do not differ greatly from each other. However, as kT/λ is reduced, $\bar{\mu}$ for $v = 10$ falls only slowly with temperature whilst for $v = -10$ the behaviour qualitatively resembles that for $v = 0$. This behaviour of $\bar{\mu}$ with kT/λ for these particular values of v may be rationalized by reference to Table 5.2. When $v = -10$ the levels E_1 and E_2 constitute the lowest energy state for the system, and when $k = 1\cdot0$ the first-order Zeeman coefficients for both the parallel and perpendicular directions are zero. This is very similar to the situation for the $^2T_2$ term in cubic symmetry when λ is positive (see Fig. 4.2). On the other hand, when $v = 10$ the levels E_4 and E_6 are the ground state and as they have non-zero first-order Zeeman coefficients associated with them, for both the parallel and perpendicular directions a sharp decrease in $\bar{\mu}$ with decreasing kT/λ would not be expected.

When λ is negative the values of $\bar{\mu}$ for $v = \pm10$ are lower than when $v = 0$ at the higher values of kT/λ, and generally show less temperature variation. In each of these two extreme examples the lowest energy states are always E_3 and E_5, which have non-zero first-order Zeeman coefficients associated with them. This observation qualitatively accounts for the difference in behaviour of $\bar{\mu}$ with $|kT/\lambda|$ when λ is negative and when λ is positive and $v > 0$.

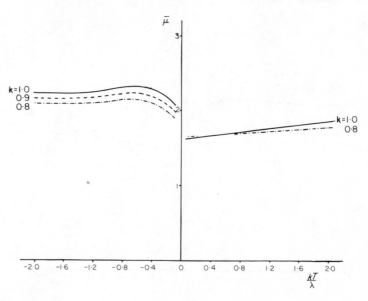

Figure 5.3. Diagrams of $\bar{\mu}$ vs. kT/λ for the $^2T_2$ term for various values of **k**, with $v = 10$.

(b) *The variation of $\bar{\mu}$ with* **k** (illustrated in Fig. 5.3). The general effect of reducing **k** below unity is that $\bar{\mu}$ tends towards the spin-only value, although the actual value of $\bar{\mu}$ attained may still be considerably different from $\mu_{s.o.}$. This trend is illustrated by the behaviour of $\bar{\mu}$ versus temperature with $v = 10$ when λ

is both positive and negative. It is also noticeable that when λ is positive the temperature variation of $\bar{\mu}$ becomes less as **k** is reduced, the higher temperature values being depressed towards 1·73 B.M. whilst the lower temperature values are raised towards this spin only value

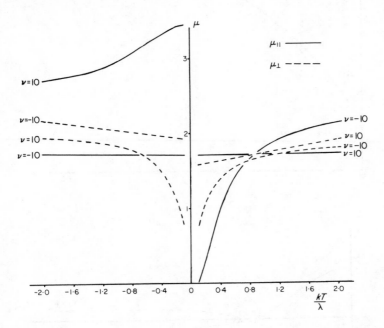

Figure 5.4. Diagrams of $\mu_{\parallel}$ and $\mu_{\perp}$ *vs.* kT/λ for the $^2T_2$ term for various values of v, with **k** = 1·0.

(c) *The variation of $\mu_{\parallel}$ and $\mu_{\perp}$ with v and kT/λ* (illustrated in Fig. 5.4). The variations of $\mu_{\parallel}$ and $\mu_{\perp}$ with kT/λ for **k** = 1·0 and $v = \pm10$ are illustrated in Fig. 5.4. Some interesting general features of distorted systems are readily illustrated with this diagram. The variation of $\mu_{\parallel}$ and $\mu_{\perp}$ with kT/λ may be quite different from that of $\bar{\mu}$, e.g. when $v = 10$ and λ is negative $\mu_{\parallel}$ increases markedly with decreasing kT/λ whilst $\mu_{\perp}$ decreases rapidly, the variation of each one being greater than for $\bar{\mu}$. A further interesting feature is also shown for both $v = \pm10$ when λ is positive and that is that there may be a particular point at which the system appears to be isotropic. In the present example this point occurs at kT/λ approximately 0·75 when $v = -10$ and at about 0·9 when $v = 10$.

The importance of measuring the magnetic anisotropy is shown in Fig. 5.5(a) and (b). The variation of $\bar{\mu}$ against kT/λ for $v = \pm1·0$ is shown in Fig. 5.5(a). It is obvious that a distinction between the situations of $v = 1·0$ and $v = -1·0$ would be difficult by merely measuring $\bar{\mu}$ unless it could be done over the range kT/λ of 0 to 0·5. For transition metal ions with small values of λ this would

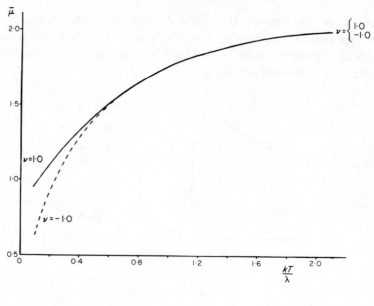

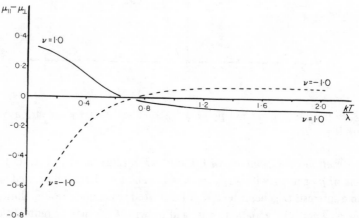

Figure 5.5. (a) A diagram of $\bar{\mu}$ vs. kT/λ for the $^2T_2$ term for $v = \pm1\cdot0$, with **k** = $1\cdot0$. (b) A diagram of the anisotropy $(\mu_\| - \mu_\perp)$ vs. kT/λ for the $^2T_2$ term for $v = \pm1\cdot0$, with **k** = $1\cdot0$.

correspond to temperatures below that attainable using liquid nitrogen. However, measurement of the anisotropy in μ would clearly distinguish between the two situations since the sign of the anisotropy is, except for $kT/\lambda = 0\cdot7$, always opposite for these particular values of v, see Fig. 5.5(b).

5.2.2. THE MAGNETIC PROPERTIES OF $^3T_1$ TERMS IN AXIALLY SYMMETRIC CRYSTAL FIELDS [4]

In this section the calculation of the magnetic properties of the $^3T_1$ term in an axially symmetric crystal field will be outlined making the following assumptions:

(a) the mixing between the $^3T_1(P)$ and $^3T_1(F)$ terms in the cubic crystal field will be allowed for (see Chapter 2).

(b) Spin–orbit coupling between the components of the $^3T_1$ ground term and the $^3T_2$ and $^3A_2$ terms will not be considered.

(c) the second-order Zeeman effect between the components of the $^3T_1$ term and the $^3T_2$ and $^3A_2$ terms will be neglected.

(d) any ligand contributions to the wave-functions will be allowed for by introducing an isotropic orbital reduction factor **k**, as in the case of the $^2T_2$ calculation.

The orbital wave-functions for the $^3T_1$ ground term after allowing for $^3T_1(P)$ $- \, ^3T_1(F)$ mixing are given in Chapter 2. However, as we are here only concerned with the contribution to the magnetic susceptibility from the $^3T_1$ term unaltered by any possible contributions from the higher $^3T_2$ and $^3A_2$ terms, the following form and abbreviations for the wave-functions appropriate to a tetragonal distortion will be most useful [4]:

$$|\pm A\rangle = (1 + c_i^2)^{-\frac{1}{2}}[\sqrt{\tfrac{15}{24}}|3, \pm 3\rangle - \sqrt{\tfrac{9}{24}}|3, \mp 1\rangle + c_i|1, \mp 1\rangle]$$

$$|0\rangle = (1 + c_i^2)^{-\frac{1}{2}}[|3, 0\rangle + c_i|1, 0\rangle]$$

where the wave-functions are specified in terms of $|L, M_L\rangle$ and the mixing coefficient, c_i, is defined in Chapter 2. The above abbreviations for the wave-functions are used since it can be shown that

$$\langle \pm A|\hat{L}_z|\pm A\rangle = \pm A \qquad \text{and} \qquad \langle \pm A|L_x|0\rangle = -\sqrt{\tfrac{1}{2}}A$$

where $A = (1 \cdot 5 - c_i^2)/(1 + c_i^2)$, all other matrix elements of this type being zero.

Each of the above orbital wave-functions have to be combined with the spin wave-functions $|M_s\rangle$, i.e. $|1\rangle$, $|0\rangle$, $|-1\rangle$ for the $^3T_1$ term.

The method of calculation is entirely the same as in the $^2T_2$ case, the difference now being that we have nine-fold degeneracy to deal with. This results in rather more computation. The secular determinant relevant to the application of a simultaneous perturbation by spin–orbit coupling and a tetragonal crystal field components may be arranged so that it factorizes into the following sub-determinants:

$$
\begin{array}{c}
\quad\quad\quad |\pm A, \mp 1\rangle \\
\langle \pm A, \mp 1| \quad \left| \left(\dfrac{\Delta}{3} - A\lambda\right) - E \right| = 0
\end{array}
$$

$$
\begin{array}{c|cc}
 & |\pm A, 0\rangle & |0, \mp 1\rangle \\
\hline
\langle \pm A, 0| & \dfrac{\Delta}{3} - E & -A\lambda \\
\langle 0, \mp 1| & -A\lambda & \dfrac{-2\Delta}{3} - E
\end{array} = 0.
$$

$$
\begin{array}{c|ccc}
 & |A, 1\rangle & |0, 0\rangle & |-A, -1\rangle \\
\hline
\langle A, 1| & \left(\dfrac{\Delta}{3} + A\lambda\right) - E & -A\lambda & 0 \\
\langle 0, 0| & -A\lambda & -\dfrac{2\Delta}{3} - E & -A\lambda \\
\langle -A, -1| & 0 & -A\lambda & \left(\dfrac{\Delta}{3} + A\lambda\right) - E
\end{array} = 0. \quad (5.22)
$$

The energies of levels after perturbation obtained from these determinants are:

$$
\frac{\Delta}{3} - A\lambda \quad \text{doubly degenerate,}
$$

$$
\frac{1}{2}\left[-\frac{\Delta}{3} + A\lambda \pm (\Delta^2 + 2A\lambda\Delta + 9A^2\lambda^2)^{\frac{1}{2}} \right],
$$

$$
\frac{\Delta}{3} + A\lambda,
$$

$$
\frac{1}{2}\left[-\frac{\Delta}{3} \pm (\Delta^2 + 4A^2\lambda^2)^{\frac{1}{2}} \right] \quad \text{each doubly degenerate.}
$$

It is fortunate in this particular case that the cubic equation factorizes and closed form expressions can be obtained for the energies. When this is not the case, and indeed when many different values of the parameters are considered, it is usual to perform the whole calculation numerically on a computer, following

Table 5.3. *The values of* $\bar{\mu}$ *for the* $^3T_1$ *term as a function of* kT/λ *(values at the top of the columns),* v *(values in the column at the left),* **k**, *and A. (Reproduced with permission from reference 4.)*

The magnetic moment for the $^3T_{1g}$ term as a function of the parameters kT/λ (values at top of columns), v (values in column at left), k, and A

$A = 1.5,\ k = 1.0$

v	0.1	0.2	0.3	0.5	0.75	1	1.5	2	−0.1	−0.2	−0.3	−0.5	−0.75	−1	−1.5	−2
10	2.18	2.36	2.42	2.48	2.53	2.58	2.66	2.74	2.62	3.30	3.58	3.81	3.89	3.89	3.86	3.83
5	1.71	1.99	2.09	2.20	2.30	2.39	2.56	2.71	2.16	2.96	3.39	3.78	3.94	3.99	4.00	3.97
4	1.57	1.85	1.97	2.10	2.22	2.33	2.53	2.70	2.07	2.86	3.35	3.77	3.96	4.02	4.04	4.01
3	1.42	1.69	1.81	1.96	2.11	2.25	2.50	2 69	1.97	2.76	3.26	3.76	3.99	4.06	4.08	4.04
2	1.25	1.49	1.61	1.79	1.99	2.17	2.47	2.68	1.89	2.66	3.18	3.75	4.01	4.10	4.12	4.08
1	1.07	1.25	1.39	1.63	1.89	2.11	2.45	2.68	1.83	2.58	3.12	3.74	4.04	4.14	4.16	4.10
0	0.89	1.09	1.27	1.56	1.85	2.09	2.44	2.68	1.83	2.55	3.10	3.74	4.05	4.16	4.17	4.11
−1	0.91	1.15	1.32	1.59	1.87	2.11	2.45	2.68	1.83	2.59	3.13	3.74	4.03	4.14	4.15	4.10
−2	0.90	1.16	1.36	1.64	1.91	2.13	2.46	2.68	1.91	2.68	3.20	3.73	3.97	4.06	4.10	4.07
−3	0.89	1.16	1.37	1.67	1.95	2.16	2.47	2.68	2.02	2.81	3.28	3.69	3.88	3.96	4.01	4.01
−4	0.89	1.16	1.37	1.68	1.97	2.18	2.48	2.69	2.14	2.94	3.34	3.64	3.77	3.84	3.91	3.93
−5	0.89	1.16	1.37	1.69	1.98	2.20	2.49	2.69	2.28	3.04	3.36	3.57	3.66	3.72	3.81	3.84
−10	0.89	1.15	1.37	1.70	2.01	2.23	2.52	2.70	2.75	3.15	3.24	3.29	3.33	3.36	3.42	3.48

$A = 1.5,\ k = 0.8$

v	0.1	0.2	0.3	0.5	0.75	1	1.5	2	−0.1	−0.2	−0.3	−0.5	−0.75	−1	−1.5	−2
10	2.26	2.43	2.48	2.52	2.56	2.59	2.65	2.70	2.40	3.04	3.30	3.52	3.61	3.62	3.60	3.57
5	1.84	2.10	2.18	2.27	2.34	2.40	2.52	2.63	1.99	2.72	3.12	3.49	3.65	3.69	3.70	3.68
4	1.71	1.98	2.07	2.17	2.26	2.33	2.48	2.61	1.90	2.63	3.06	3.48	3.66	3.72	3.73	3.71
3	1.58	1.83	1.93	2.04	2.15	2.25	2.44	2.59	1.81	2.53	2.99	3.46	3.67	3.75	3.77	3.74
2	1.44	1.65	1.75	1.87	2.02	2.16	2.40	2.58	1.73	2.44	2.92	3.44	3.69	3.78	3.80	3.76
1	1.29	1.43	1.53	1.71	1.91	2.09	2.38	2.57	1.67	2.36	2.86	3.43	3.71	3.80	3.83	3.78
0	1.14	1.28	1.41	1.63	1.87	2.07	2.37	2.56	1.65	2.34	2.83	3.42	3.71	3.81	3.84	3.79
−1	1.13	1.33	1.46	1.67	1.90	2.09	2.37	2.57	1.68	2.37	2.86	3.43	3.70	3.80	3.82	3.78
−2	1.11	1.34	1.50	1.73	1.94	2.12	2.39	2.57	1.75	2.47	2.94	3.43	3.66	3.75	3.78	3.75
−3	1.10	1.33	1.51	1.76	1.98	2.15	2.41	2.58	1.87	2.60	3.04	3.43	3.59	3.67	3.72	3.71
−4	1.10	1.33	1.51	1.78	2.01	2.18	2.42	2.59	2.00	2.74	3.12	3.40	3.52	3.58	3.64	3.65
−5	1.10	1.32	1.51	1.78	2.02	2.20	2.44	2.60	2.13	2.86	3.16	3.36	3.45	3.50	3.56	3.58
−10	1.09	1.32	1.51	1.80	2.05	2.24	2.47	2.62	2.64	3.04	3.13	3.18	3.20	3.23	3.27	3.31

$A = 1.5,\ k = 0.6$

v	0.1	0.2	0.3	0.5	0.75	1	1.5	2	−0.1	−0.2	−0.3	−0.5	−0.75	−1	−1.5	−2
10	2.34	2.50	2.55	2.58	2.60	2.62	2.65	2.69	2.19	2.78	3.23	3.24	3.34	3.36	3.35	3.33
5	1.97	2.22	2.30	2.36	2.41	2.45	2.52	2.59	1.81	2.48	2.85	3.20	3.35	3.41	3.42	3.41
4	1.87	2.12	2.20	2.27	2.33	2.38	2.48	2.57	1.73	2.39	2.79	3.18	3.36	3.42	3.44	3.43
3	1.76	2.00	2.08	2.16	2.23	2.30	2.43	2.55	1.65	2.30	2.73	3.16	3.36	3.44	3.47	3.45
2	1.65	1.84	1.92	2.01	2.11	2.21	2.39	2.53	1.57	2.21	2.65	3.14	3.37	3.46	3.49	3.47
1	1.54	1.66	1.73	1.86	2.01	2.15	2.37	2.52	1.52	2.14	2.59	3.12	3.37	3.47	3.51	3.48
0	1.45	1.54	1.63	1.79	1.97	2.13	2.36	2.52	1.50	2.12	2.57	3.11	3.38	3.48	3.52	3.49
−1	1.40	1.57	1.67	1.83	2.00	2.14	2.37	2.52	1.52	2.15	2.60	3.12	3.38	3.47	3.51	3.48
−2	1.37	1.56	1.70	1.88	2.04	2.18	2.38	2.53	1.60	2.25	2.69	3.14	3.35	3.44	3.48	3.46
−3	1.35	1.55	1.70	1.91	2.08	2.21	2.40	2.54	1.71	2.39	2.80	3.16	3.32	3.39	3.43	3.43
−4	1.34	1.54	1.70	1.92	2.10	2.23	2.42	2.55	1.85	2.55	2.90	3.17	3.28	3.33	3.38	3.39
−5	1.34	1.54	1.70	1.93	2.12	2.25	2.44	2.56	2.00	2.68	2.97	3.16	3.23	3.28	3.32	3.34
−10	1.33	1.52	1.69	1.94	2.15	2.29	2.48	2.59	2.55	2.93	3.02	3.06	3.09	3.10	3.13	3.16

$A = 1.5,\ k = 0.4$

v	0.1	0.2	0.3	0.5	0.75	1	1.5	2	−0.1	−0.2	−0.3	−0.5	−0.75	−1	−1.5	−2
10	2.42	2.58	2.62	2.64	2.66	2.67	2.68	2.70	1.98	2.52	2.76	2.97	3.08	3.11	3.13	3.12
5	2.11	2.36	2.42	2.47	2.50	2.52	2.57	2.61	1.64	2.24	2.59	2.92	3.07	3.13	3.16	3.16
4	2.02	2.28	2.34	2.40	2.43	2.47	2.53	2.59	1.56	2.16	2.53	2.90	3.07	3.14	3.17	3.17
3	1.94	2.18	2.25	2.30	2.35	2.40	2.49	2.57	1.48	2.08	2.46	2.87	3.07	3.14	3.19	3.18
2	1.87	2.06	2.12	2.19	2.26	2.33	2.45	2.55	1.41	1.99	2.39	2.84	3.06	3.15	3.20	3.19
1	1.81	1.92	1.97	2.07	2.18	2.27	2.43	2.54	1.36	1.93	2.33	2.81	3.06	3.16	3.21	3.20
0	1.78	1.84	1.90	2.02	2.15	2.26	2.42	2.54	1.34	1.90	2.30	2.80	3.06	3.16	3.21	3.20
−1	1.69	1.84	1.93	2.04	2.17	2.27	2.43	2.54	1.37	1.93	2.34	2.81	3.06	3.16	3.21	3.20
−2	1.64	1.82	1.94	2.08	2.20	2.30	2.44	2.55	1.44	2.03	2.44	2.86	3.06	3.14	3.19	3.19
−3	1.62	1.80	1.94	2.10	2.23	2.32	2.46	2.56	1.56	2.19	2.57	2.91	3.06	3.12	3.17	3.17
−4	1.61	1.79	1.93	2.11	2.25	2.34	2.48	2.57	1.71	2.35	2.69	2.94	3.05	3.10	3.14	3.15
−5	1.60	1.78	1.92	2.11	2.26	2.36	2.49	2.58	1.86	2.50	2.78	2.96	3.04	3.07	3.11	3.12
−10	1.59	1.76	1.91	2.12.	2.28	2.40	2.53	2.61	2.45	2.83	2.91	2.96	2.98	2.99	3.00	3.02

$A = 1.3,\ k = 1.0$

v	0.1	0.2	0.3	0.5	0.75	1	1.5	2	−0.1	−0.2	−0.3	−0.5	−0.75	−1	−1.5	−2
10	2.36	2.48	2.52	2.56	2.60	2.63	2.70	2.76	2.72	3.28	3.49	3.65	3.70	3.69	3.65	3.62
5	1.95	2.17	2.25	2.33	2.41	2.48	2.60	2.72	2.27	3.01	3.37	3.66	3.76	3.78	3.77	3.73
4	1.82	2.05	2.14	2.24	2.33	2.42	2.57	2.71	2.16	2.92	3.32	3.66	3.78	3.81	3.80	3.77
3	1.66	1.90	1.99	2.11	2.23	2.34	2.54	2.70	2.04	2.82	3.26	3.65	3.81	3.85	3.84	3.80
2	1.48	1.69	1.80	1.94	2.11	2.26	2.52	2.69	1.94	2.71	3.19	3.65	3.84	3.89	3.88	3.83
1	1.27	1.44	1.55	1.76	2.00	2.20	2.50	2.69	1.86	2.62	3.13	3.65	3.87	3.93	3.91	3.85
0	1.08	1.26	1.41	1.68	1.96	2.18	2.49	2.69	1.83	2.58	3.10	3.65	3.89	3.95	3.93	3.86
−1	1.09	1.32	1.47	1.75	1.99	2.20	2.50	2.69	1.87	2.63	3.14	3.65	3.86	3.93	3.91	3.85
−2	1.07	1.33	1.52	1.78	2.03	2.23	2.51	2.69	1.97	2.75	3.21	3.62	3.79	3.85	3.86	3.82
−3	1.07	1.33	1.53	1.81	2.07	2.25	2.52	2.70	2.12	2.90	3.28		3.69	3.74	3.78	3.77
−4	1.06	1.33	1.53	1.83	2.09	2.28	2.53	2.70	2.28	3.01	3.30	3.49	3.58	3.63	3.69	3.70
−5	1.06	1.33	1.53	1.84	2.11	2.29	2.54	2.70	2.44	3.08	3.29	3.42	3.48	3.53	3.60	3.63
−10	1.06	1.32	1.53	1.86	2.14	2.33	2.57	2.89	2.85	3.09	3.14	3.18	3.21	3.23	3.28	3.33

Table 5.3 (continued).

$A = 1\cdot3,\ k = 0\cdot8$

	0·1	0·2	0·3	0·5	0·75	1	1·5	2		−0·1	−0·2	−0·3	−0·5	−0·75	−1	−1·5	−2
10	2·42	2·53	2·57	2·60	2·62	2·65	2·69	2·73		2·52	3·04	3·25	3·41	3·47	3·47	3·44	3·41
5	2·06	2·26	2·33	2·39	2·44	2·49	2·58	2·66		2·10	2·79	3·13	3·41	3·51	3·54	3·53	3·50
4	1·94	2·16	2·23	2·30	2·37	2·43	2·54	2·65		1·99	2·71	3·08	3·40	3·52	3·56	3·55	3·52
3	1·80	2·02	2·10	2·18	2·27	2·35	2·51	2·63		1·89	2·61	3·02	3·39	3·54	3·59	3·58	3·55
2	1·65	1·84	1·92	2·03	2·15	2·27	2·47	2·62		1·79	2·50	2·95	3·38	3·56	3·62	3·61	3·57
1	1·47	1·61	1·69	1·85	2·04	2·21	2·45	2·61		1·72	2·42	2·89	3·38	3·58	3·65	3·64	3·59
0	1·32	1·45	1·57	1·78	2·00	2·18	2·44	2·61		1·69	2·38	2·86	3·37	3·59	3·66	3·65	3·60
−1	1·30	1·50	1·62	1·82	2·03	2·20	2·45	2·61		1·72	2·43	2·90	3·37	3·58	3·64	3·64	3·59
−2	1·28	1·50	1·66	1·88	2·07	2·23	2·46	2·62		1·83	2·55	2·98	3·36	3·53	3·59	3·60	3·56
−3	1·26	1·50	1·67	1·91	2·11	2·26	2·48	2·63		1·98	2·71	3·70	3·34	3·45	3·51	3·54	3·52
−4	1·26	1·49	1·67	1·93	2·14	2·29	2·49	2·63		2·15	2·85	3·12	3·30	3·38	3·42	3·47	3·47
−5	1·26	1·49	1·67	1·93	2·15	2·31	2·51	2·64		2·31	2·94	3·14	3·26	3·31	3·35	3·40	3·42
−10	1·25	1·48	1·67	1·95	2·19	2·35	2·54	2·66		2·77	3·01	3·06	3·09	3·11	3·13	3·16	3·20

$A = 1\cdot3,\ k = 0\cdot6$

	0·1	0·2	0·3	0·5	0·75	1	1·5	2		−0·1	−0·2	−0·3	−0·5	−0·75	−1	−1·5	−2
10	2·48	2·59	2·62	2·64	2·66	2·67	2·69	2·72		2·32	2·81	3·01	3·18	3·25	3·26	3·24	3·22
5	2·16	2·36	2·41	2·46	2·50	2·53	2·59	2·64		1·93	2·57	2·89	3·16	3·27	3·30	3·30	3·28
4	2·06	2·27	2·33	2·39	2·43	2·47	2·55	2·62		1·83	2·49	2·84	3·15	3·27	3·31	3·32	3·30
3	1·94	2·15	2·22	2·28	2·34	2·40	2·51	2·61		1·74	2·40	-2·78	3·13	3·28	3·33	3·34	3·32
2	1·82	2·00	2·07	2·15	2·23	2·32	2·48	2·59		1·64	2·30	2·71	3·12	3·29	3·35	3·36	3·33
1	1·69	1·81	1·87	1·99	2·14	2·26	2·46	2·58		1·57	2·21	2·65	3·10	3·30	3·37	3·38	3·35
0	1·59	1·68	1·77	1·93	2·10	2·24	2·45	2·58		1·54	2·18	2·62	3·09	3·31	3·38	3·39	3·35
−1	1·54	1·71	1·81	1·96	2·13	2·26	2·45	2·58		1·58	2·22	2·66	3·10	3·30	3·37	3·38	3·34
−2	1·50	1·71	1·84	2·01	2·17	2·29	2·47	2·59		1·69	2·35	2·76	3·12	3·27	3·33	3·35	3·33
−3	1·48	1·69	1·85	2·04	2·20	2·32	2·48	2·60		1·84	2·53	2·87	3·12	3·23	3·28	3·31	3·30
−4	1·48	1·69	1·84	2·06	2·22	2·34	2·50	2·60		2·02	2·68	2·94	3·12	3·19	3·23	3·26	3·27
−5	1·47	1·68	1·84	2·06	2·24	2·36	2·51	2·61		2·19	2·79	2·98	3·10	3·15	3·18	3·21	3·23
−10	1·46	1·67	1·83	2·07	2·27	2·40	2·55	2·64		2·69	2·93	2·98	3·01	3·02	3·04	3·06	3·08

$A = 1\cdot3,\ k = 0\cdot4$

	0·1	0·2	0·3	0·5	0·75	1	1·5	2		−0·1	−0·2	−0·3	−0·5	−0·75	−1	−1·5	−2
10	2·54	2·65	2·67	2·69	2·70	2·70	2·72	2·73		2·12	2·58	2·77	2·95	3·03	3·06	3·06	3·05
5	2·27	2·47	2·51	2·55	2·57	2·59	2·62	2·66		1·76	2·36	2·65	2·92	3·03	3·08	3·09	3·09
4	2·19	2·39	2·44	2·49	2·52	2·54	2·59	2·64		1·67	2·28	2·60	2·90	3·03	3·08	3·10	3·10
3	2·10	2·30	2·35	2·40	2·44	2·48	2·56	2·62		1·58	2·19	2·54	2·88	3·03	3·09	3·11	3·11
2	2·00	2·18	2·23	2·29	2·36	2·42	2·53	2·61		1·50	2·09	2·48	2·86	3·03	3·10	3·13	3·12
1	1·92	2·03	2·08	2·17	2·28	2·37	2·51	2·60		1·43	2·01	2·41	2·83	3·03	3·10	3·14	3·12
0	1·88	1·94	2·01	2·12	2·25	2·35	2·50	2·60		1·40	1·97	2·38	2·82	3·03	3·11	3·14	3·13
−1	1·79	1·95	2·03	2·15	2·27	2·37	2·51	2·60		1·43	2·02	2·42	2·84	3·03	3·10	3·13	3·12
−2	1·74	1·93	2·05	2·19	2·30	2·39	2·52	2·61		1·54	2·16	2·53	2·88	3·03	3·09	3·12	3·11
−3	1·72	1·91	2·05	2·21	2·33	2·42	2·54	2·61		1·71	2·34	2·67	2·91	3·02	3·06	3·10	3·10
−4	1·71	1·90	2·04	2·22	2·35	2·44	2·55	2·62		1·89	2·52	2·77	2·94	3·01	3·04	3·07	3·08
−5	1·70	1·89	2·03	2·22	2·36	2·45	2·56	2·63		2·07	2·64	2·84	2·95	2·99	3·02	3·05	3·06
−10	1·69	1·87	2·02	2·23	2·38	2·48	2·59	2·66		2·62	2·86	2·90	2·93	2·94	2·95	2·96	2·97

$A = 1\cdot2,\ k = 1\cdot0$

	0·1	0·2	0·3	0·5	0·75	1	1·5	2		−0·1	−0·2	−0·3	−0·5	−0·75	−1	−1·5	−2
10	2·44	2·53	2·56	2·60	2·63	2·66	2·71	2·76		2·77	3·25	3·44	3·57	3·60	3·59	3·55	3·51
5	2·07	2·26	2·32	2·39	2·46	2·52	2·63	2·73		2·33	3·03	3·35	3·59	3·66	3·68	3·65	3·62
4	1·94	2·15	2·22	2·31	2·39	2·46	2·60	2·72		2·21	2·95	3·31	3·59	3·69	3·71	3·69	3·65
3	1·79	2·00	2·08	2·19	2·29	2·39	2·57	2·71		2·09	2·85	3·26	3·59	3·71	3·74	3·72	3·68
2	1·60	1·80	1·89	2·03	2·18	2·32	2·55	2·70		1·98	2·74	3·20	3·60	3·78	3·76	3·71	3·61
1	1·39	1·54	1·64	1·84	2·07	2·26	2·53	2·70		1·89	2·65	3·14	3·60	3·78	3·82	3·79	3·73
0	1·18	1·35	1·50	1·76	2·03	2·23	2·52	2·70		1·85	2·60	3·11	3·60	3·80	3·84	3·80	3·74
−1	1·19	1·41	1·56	1·80	2·05	2·25	2·53	2·70		1·89	2·66	3·14	3·59	3·77	3·82	3·79	3·73
−2	1·17	1·43	1·61	1·86	2·10	2·28	2·54	2·70		2·02	2·79	3·21	3·55	3·69	3·74	3·74	3·70
−3	1·16	1·43	1·62	1·90	2·13	2·31	2·55	2·71		2·19	2·94	3·26	3·49	3·59	3·64	3·67	3·65
−4	1·16	1·42	1·63	1·91	2·16	2·33	2·56	2·71		2·37	3·04	3·27	3·41	3·48	3·53	3·58	3·59
−5	1·15	1·42	1·63	1·92	2·17	2·35	2·57	2·71		2·53	3·09	3·25	3·34	3·40	3·44	3·50	3·52
−10	1·15	1·41	1·63	1·94	2·20	2·38	2·60	2·73		2·89	3·06	3·10	3·13	3·15	3·17	3·22	3·26

$A = 1\cdot2,\ k = 0\cdot8$

	0·1	0·2	0·3	0·5	0·75	1	1·5	2		−0·1	−0·2	−0·3	−0·5	−0·75	−1	−1·5	−2
10	2·49	2·58	2·60	2·63	2·65	2·67	2·71	2·74		2·58	3·03	3·21	3·36	3·40	3·39	3·36	3·33
5	2·16	2·34	2·39	2·44	2·49	2·53	2·61	2·68		2·17	2·82	3·12	3·36	3·44	3·46	3·44	3·41
4	2·05	2·24	2·30	2·37	2·42	2·48	2·58	2·67		2·06	2·75	3·08	3·36	3·45	3·48	3·47	3·44
3	1·91	2·11	2·18	2·26	2·33	2·41	2·54	2·65		1·94	2·65	3·03	3·35	3·47	3·51	3·49	3·46
2	1·75	1·93	2·01	2·11	2·22	2·33	2·51	2·64		1·83	2·55	2·97	3·35	3·50	3·54	3·52	3·48
1	1·57	1·70	1·78	1·93	2·11	2·27	2·49	2·64		1·75	2·45	2·91	3·34	3·52	3·56	3·55	3·50
0	1·41	1·54	1·65	1·86	2·07	2·25	2·49	2·63		1·71	2·41	2·88	3·34	3·53	3·58	3·56	3·50
−1	1·39	1·58	1·71	1·90	2·10	2·26	2·49	2·63		1·75	2·46	2·91	3·34	3·51	3·56	3·54	3·49
−2	1·36	1·59	1·75	1·95	2·14	2·29	2·50	2·64		1·88	2·60	3·00	3·33	3·46	3·50	3·51	3·47
−3	1·35	1·59	1·76	1·99	2·18	2·32	2·52	2·65		2·06	2·77	3·08	3·29	3·38	3·43	3·45	3·44
−4	1·34	1·58	1·76	2·00	2·21	2·35	2·53	2·65		2·25	2·89	3·11	3·25	3·31	3·35	3·39	3·39
−5	1·34	1·58	1·76	2·01	2·22	2·36	2·54	2·66		2·42	2·96	3·11	3·20	3·25	3·28	3·32	3·34
−10	1·33	1·57	1·76	2·03	2·25	2·40	2·58	2·68		2·82	3·00	3·03	3·05	3·07	3·09	3·12	3·15

Table 5.3 (continued).

$A = 1\cdot2,\ k = 0\cdot6$

0·1	0·1	0·2	0·3	0·5	0·75	1	1·5	2	−0·1	−0·2	−0·3	−0·5	−0·75	−1	−1·5	−2
10	2·54	2·63	2·65	2·67	2·68	2·69	2·71	2·73	2·38	2·82	2·99	3·15	3·20	3·21	3·19	3·17
5	2·26	2·43	2·47	2·51	2·54	2·57	2·62	2·67	2·00	2·62	2·90	3·13	3·22	3·25	3·24	3·22
4	2·16	2·34	2·39	2·44	2·48	2·51	2·58	2·65	1·90	2·54	2·86	3·13	3·23	3·26	3·26	3·24
3	2·04	2·23	2·29	2·35	2·40	2·45	2·55	2·63	1·79	2·45	2·81	3·12	3·24	3·28	3·28	3·26
2	1·91	2·08	2·14	2·21	2·30	2·38	2·52	2·62	1·69	2·35	2·74	3·10	3·25	3·30	3·30	3·27
1	1·77	1·89	1·95	2·06	2·20	2·32	2·50	2·61	1·61	2·26	2·68	3·09	3·26	3·31	3·31	3·28
0	1·66	1·76	1·84	2·00	2·17	2·30	2·49	2·61	1·57	2·21	2·65	3·08	3·27	3·32	3·32	3·28
−1	1·61	1·78	1·88	2·04	2·19	2·32	2·50	2·61	1·61	2·27	2·69	3·09	3·26	3·31	3·31	3·28
−2	1·57	1·78	1·91	2·08	2·23	2·35	2·51	2·62	1·74	2·42	2·79	3·10	3·23	3·28	3·29	3·26
−3	1·55	1·77	1·92	2·11	2·26	2·37	2·53	2·62	1·93	2·60	2·89	3·10	3·19	3·23	3·25	3·24
−4	1·55	1·76	1·92	2·13	2·29	2·40	2·54	2·63	2·13	2·74	2·95	3·09	3·14	3·18	3·21	3·21
−5	1·54	1·75	1·92	2·13	2·30	2·41	2·55	2·64	2·31	2·83	2·98	3·07	3·11	3·13	3·16	3·17
−10	1·53	1·74	1·91	2·14	2·33	2·45	2·59	2·67	2·75	2·93	2·96	2·98	3·00	3·01	3·02	3·04

$A = 1\cdot2,\ k = 0\cdot4$

	0·1	0·2	0·3	0·5	0·75	1	1·5	2	−0·1	−0·2	−0·3	−0·5	−0·75	−1	−1·5	−2
10	2·59	2·68	2·69	2·71	2·72	2·72	2·73	2·74	2·19	2·60	2·78	2·94	3·01	3·03	3·03	3·02
5	2·36	2·52	2·55	2·58	2·60	2·62	2·65	2·68	1·84	2·41	2·68	2·91	3·01	3·05	3·06	3·05
4	2·27	2·45	2·49	2·53	2·55	2·58	2·62	2·67	1·75	2·34	2·64	2·90	3·01	3·05	3·07	3·06
3	2·18	2·36	2·41	2·45	2·49	2·52	2·59	2·65	1·64	2·25	2·59	2·89	3·01	3·06	3·08	3·07
2	2·08	2·24	2·29	2·35	2·41	2·47	2·57	2·64	1·55	2·15	2·52	2·86	3·01	3·07	3·09	3·08
1	1·98	2·09	2·14	2·23	2·33	2·42	2·55	2·63	1·47	2·06	2·45	2·84	3·02	3·08	3·10	3·09
0	1·93	2·00	2·06	2·18	2·30	2·40	2·54	2·63	1·43	2·02	2·42	2·83	3·02	3·08	3·10	3·09
−1	1·84	2·00	2·09	2·21	2·32	2·42	2·55	2·63	1·48	2·07	2·46	2·85	3·01	3·08	3·10	3·08
−2	1·79	1·99	2·10	2·24	2·35	2·44	2·56	2·63	1·61	2·23	2·59	2·88	3·01	3·06	3·08	3·08
−3	1·77	1·97	2·10	2·26	2·38	2·46	2·57	2·64	1·80	2·43	2·71	2·91	3·00	3·04	3·06	3·06
−4	1·76	1·96	2·10	2·27	2·40	2·48	2·58	2·65	2·01	2·60	2·80	2·93	2·99	3·01	3·04	3·04
−5	1·75	1·95	2·09	2·28	2·41	2·49	2·59	2·66	2·20	2·71	2·85	2·94	2·97	2·99	3·02	3·03
−10	1·74	1·93	2·08	2·28	2·43	2·52	2·63	2·68	2·68	2·86	2·90	2·92	2·92	2·93	2·94	2·95

$A = 1\cdot0,\ k = 1\cdot0$

	0·1	0·2	0·3	0·5	0·75	1	1·5	2	−0·1	−0·2	−0·3	−0·5	−0·75	−1	−1·5	−2
10	2·57	2·63	2·65	2·67	2·69	2·71	2·75	2·78	2·83	3·18	3·32	3·41	3·42	3·40	3·35	3·32
5	2·30	2·42	2·46	2·51	2·56	2·60	2·68	2·75	2·48	3·32	3·28	3·44	3·47	3·47	3·44	3·41
4	2·19	2·34	2·38	2·44	2·50	2·55	2·65	2·74	2·36	2·99	3·26	3·44	3·49	3·49	3·46	3·43
3	2·05	2·21	2·27	2·34	2·42	2·49	2·63	2·74	2·22	2·91	3·23	3·45	3·52	3·53	3·49	3·45
2	1·86	2·03	2·10	2·20	2·32	2·43	2·61	2·73	2·08	2·81	3·18	3·47	3·56	3·57	3·53	3·48
1	1·62	1·76	1·85	2·02	2·21	2·37	2·59	2·72	1·95	2·71	3·14	3·48	3·59	3·60	3·55	3·49
0	1·40	1·55	1·69	1·93	2·17	2·35	2·59	2·72	1·90	2·66	3·11	3·49	3·61	3·62	3·56	3·50
−1	1·40	1·62	1·76	1·98	2·20	2·37	2·59	2·72	1·97	2·72	3·14	3·47	3·58	3·60	3·55	3·49
−2	1·38	1·63	1·81	2·04	2·24	2·40	2·60	2·73	2·16	2·88	3·19	3·41	3·50	3·53	3·51	3·47
−3	1·37	1·63	1·82	2·07	2·28	2·42	2·61	2·73	2·39	3·01	3·20	3·33	3·40	3·43	3·45	3·43
−4	1·36	1·63	1·83	2·09	2·30	2·44	2·62	2·73	2·59	3·06	3·18	3·26	3·31	3·34	3·38	3·38
−5	1·36	1·63	1·83	2·10	2·32	2·46	2·63	2·74	2·72	3·07	3·15	3·20	3·24	3·27	3·31	3·33
−10	1·35	1·62	1·83	2·12	2·35	2·49	2·65	2·75	2·92	3·00	3·02	3·04	3·05	3·07	3·10	3·13

$A = 1\cdot0,\ k = 0\cdot8$

	0·1	0·2	0·3	0·5	0·75	1	1·5	2	−0·1	−0·2	−0·3	−0·5	−0·75	−1	−1·5	−2
10	2·61	2·66	2·67	2·69	2·70	2·72	2·74	2·77	2·66	3·00	3·14	3·24	3·26	3·25	3·22	3·19
5	2·37	2·48	2·52	2·55	2·58	2·61	2·67	2·72	2·32	2·87	3·09	3·25	3·30	3·30	3·28	3·25
4	2·27	2·41	2·45	2·49	2·53	2·57	2·64	2·71	2·21	2·81	3·07	3·26	3·31	3·32	3·30	3·27
3	2·14	2·29	2·34	2·40	2·46	2·51	2·62	2·70	2·08	2·74	3·04	3·26	3·33	3·34	3·32	3·29
2	1·98	2·13	2·19	2·27	2·36	2·45	2·59	2·69	1·94	2·64	2·99	3·26	3·36	3·37	3·34	3·31
1	1·78	1·90	1·97	2·10	2·26	2·39	2·57	2·68	1·82	2·53	2·94	3·27	3·38	3·40	3·36	3·32
0	1·60	1·72	1·83	2·02	2·22	2·37	2·57	2·68	1·77	2·48	2·91	3·27	3·40	3·41	3·37	3·32
−1	1·57	1·77	1·89	2·07	2·25	2·39	2·57	2·68	1·84	2·55	2·95	3·26	3·37	3·39	3·36	3·32
−2	1·54	1·78	1·93	2·12	2·29	2·41	2·58	2·69	2·03	2·72	3·02	3·23	3·32	3·34	3·33	3·30
−3	1·53	1·77	1·94	2·15	2·32	2·44	2·59	2·69	2·27	2·87	3·06	3·18	3·24	3·27	3·29	3·27
−4	1·52	1·77	1·94	2·17	2·35	2·46	2·61	2·70	2·48	2·95	3·07	3·24	3·18	3·21	3·23	3·21
−5	1·52	1·76	1·94	2·18	2·36	2·48	2·62	2·70	2·64	2·98	3·05	3·10	3·13	3·15	3·18	3·20
−10	1·51	1·75	1·94	2·20	2·39	2·51	2·64	2·72	2·87	2·96	2·97	2·99	3·00	3·01	3·03	3·05

$A = 1\cdot0,\ k = 0\cdot6$

	0·1	0·2	0·3	0·5	0·75	1	1·5	2	−0·1	−0·2	−0·3	−0·5	−0·75	−1	−1·5	−2
10	2·64	2·69	2·70	2·72	2·72	2·73	2·75	2·76	2·49	2·82	2·96	3·08	3·11	3·11	3·09	3·07
5	2·44	2·54	2·57	2·60	2·62	2·64	2·68	2·71	2·17	2·69	2·91	3·08	3·13	3·14	3·13	3·11
4	2·35	2·48	2·51	2·55	2·57	2·60	2·65	2·70	2·06	2·64	2·88	3·07	3·14	3·15	3·14	3·13
3	2·24	2·38	2·43	2·47	2·51	2·55	2·62	2·69	1·94	2·56	2·85	3·07	3·15	3·17	3·16	3·14
2	2·11	2·25	2·29	2·35	2·42	2·49	2·60	2·68	1·81	2·46	2·80	3·07	3·16	3·19	3·18	3·15
1	1·94	2·05	2·10	2·21	2·34	2·44	2·58	2·67	1·70	2·36	2·74	3·06	3·18	3·20	3·19	3·16
0	1·81	1·90	1·99	2·15	2·30	2·42	2·58	2·67	1·65	2·30	2·71	3·06	3·18	3·21	3·19	3·16
−1	1·76	1·94	2·04	2·18	2·33	2·44	2·58	2·67	1·71	2·38	2·75	3·06	3·17	3·20	3·19	3·16
−2	1·72	1·94	2·07	2·23	2·36	2·46	2·59	2·67	1·91	2·57	2·85	3·06	3·14	3·17	3·17	3·14
−3	1·69	1·93	2·07	2·26	2·39	2·48	2·60	2·68	2·16	2·74	2·92	3·04	3·10	3·12	3·14	3·13
−4	1·69	1·92	2·07	2·27	2·41	2·50	2·62	2·69	2·38	2·84	2·95	3·02	3·06	3·08	3·10	3·10
−5	1·69	1·91	2·07	2·28	2·43	2·52	2·63	2·69	2·55	2·89	2·96	3·00	3·03	3·05	3·07	3·08
−10	1·67	1·90	2·07	2·29	2·45	2·55	2·65	2·71	2·83	2·91	2·93	2·94	2·95	2·95	2·97	2·98

Table 5.3 (continued).

$A = 1·0, k = 0·4$

0·1	0·1	0·2	0·3	0·5	0·75	1	1·5	2	−0·1	−0·2	−0·3	−0·5	−0·75	−1	−1·5	−2
10	2·68	2·73	2·74	2·74	2·75	2·75	2·76	2·77	2·32	2·64	2·78	2·91	2·96	2·98	2·97	2·97
5	2·51	2·61	2·63	2·65	2·67	2·68	2·70	2·72	2·02	2·52	2·73	2·90	2·97	2·99	3·00	2·99
4	2·44	2·56	2·59	2·61	2·63	2·64	2·68	2·71	1·92	2·46	2·70	2·90	2·97	3·00	3·00	3·00
3	2·35	2·48	2·51	2·55	2·58	2·60	2·66	2·70	1·80	2·38	2·66	2·89	2·98	3·00	3·01	3·00
2	2·23	2·37	2·41	2·46	2·51	2·56	2·64	2·69	1·68	2·29	2·61	2·87	2·98	3·01	3·02	3·01
1	2·11	2·21	2·26	2·34	2·44	2·52	2·62	2·68	1·57	2·18	2·54	2·86	2·98	3·02	3·03	3·01
0	2·03	2·10	2·17	2·29	2·41	2·50	2·62	2·68	1·52	2·13	2·51	2·85	2·98	3·02	3·03	3·02
−1	1·95	2·12	2·20	2·32	2·43	2·51	2·62	2·68	1·59	2·21	2·56	2·86	2·98	3·02	3·03	3·01
−2	1·90	2·10	2·22	2·36	2·46	2·53	2·63	2·69	1·79	2·41	2·69	2·89	2·97	3·00	3·02	3·01
−3	1·88	2·09	2·22	2·38	2·48	2·55	2·64	2·69	2·05	2·61	2·79	2·91	2·96	2·98	3·00	3·00
−4	1·87	2·08	2·22	2·39	2·50	2·57	2·65	2·70	2·29	2·73	2·85	2·91	2·95	2·96	2·98	2·98
−5	1·86	2·07	2·22	2·39	2·51	2·58	2·66	2·70	2·46	2·80	2·87	2·91	2·93	2·95	2·96	2·97
−10	1·85	2·05	2·21	2·40	2·53	2·60	2·68	2·72	2·78	2·87	2·88	2·89	2·90	2·90	2·91	2·91

the steps outlined in the calculation for the $^2T_2$ term. This has been done for the $^3T_1$ term and tables for $\bar{\mu}$ in terms of the parameters A, k, v and kT/λ are available in the literature [4] and are summarized in Table 5.3. Some examples of the variation of $\bar{\mu}$ with kT/λ for a few values of the parameters v, k, and A are given in Figs. 5.6–5.8.

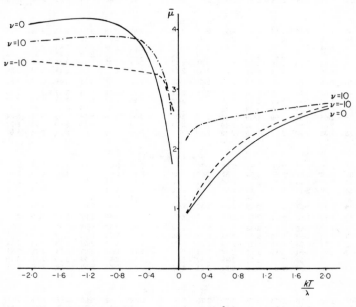

Figure 5.6. The variation of $\bar{\mu}$ vs. kT/λ for the $^3T_1$ term for various values of v, with $k = 1·0$ and $A = 1·5$.

(a) *The variation of $\bar{\mu}$ with v* (illustrated in Fig. 5.6). The result of a positive and increasing value of v is to make $\bar{\mu}$ at high values of kT/λ increase towards the spin-only value of 2·83 B.M. when λ is positive and reduce it towards the spin-only value when λ is negative. The variation of $\bar{\mu}$ with decreasing $|kT/\lambda|$ is significantly reduced compared with the undistorted situation. An increasingly negative value of v also makes $\bar{\mu}$ at high values of kT/λ tend towards the spin-only value. However, the variation of $\bar{\mu}$ with decreasing $|kT/\lambda|$ now depends on the sign of λ. When λ is negative, the variation of $\bar{\mu}$ with $|kT/\lambda|$ is reduced just as in the case of positive values of v. On the other hand, when λ is positive the degree of variation is marginally increased, the values of $\bar{\mu}$ at smaller values of kT/λ being smaller than in the cubic field situation.

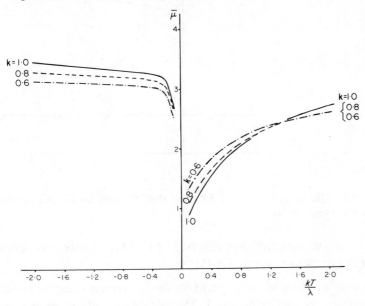

Figure 5.7. The variation of $\bar{\mu}$ vs. kT/λ for the $^3T_1$ term for various values of **k** with $v = -10$ and $A = 1·5$.

(b) *The variation of $\bar{\mu}$ with* **k** (illustrated in Fig. 5.7). At high values of kT/λ, with λ positive, a reduction in **k** reduces $\bar{\mu}$ at all values of v. On the other hand, at low values of kT/λ, $\bar{\mu}$ tends to increase towards the spin-only value thus apparently reducing the temperature variation compared with **k** $= 1·0$. When λ is negative $\bar{\mu}$ decreases with decreasing **k** for all values of v and kT/λ.

(c) *The variation of $\bar{\mu}$ with* A (illustrated in Fig. 5.8). Reducing A from the weak field value of $1·5$ to the medium field value of $1·0$, where there is maximum mixing between the components of the $^3T_1(F)$ and $^3T_1(P)$ terms, does not have the same effect on $\bar{\mu}$ for all values of v and kT/λ. At all values of v, when λ is positive there is little effect on $\bar{\mu}$ at high values of kT/λ as A is reduced. However, at low values of kT/λ, $\bar{\mu}$ is increased, thus making the temperature variation become less as A is reduced. A similar lessening of the temperature variation occurs when λ is negative. In this case this arises because $\bar{\mu}$ is reduced at high values of $|kT/\lambda|$ as A is reduced, whereas $\bar{\mu}$ is increased at low values of $|kT/\lambda|$ as A is reduced.

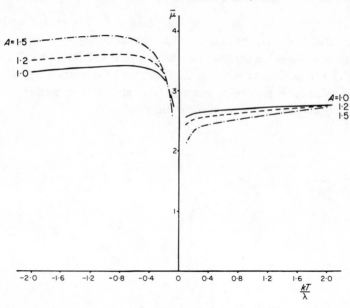

Figure 5.8. The variation of $\bar{\mu}$ *vs.* kT/λ for the $^3T_1$ term for various values of A, with $\mathbf{k} = 1 \cdot 0$ and $v = 10$.

5.2.3. THE MAGNETIC PROPERTIES OF $^5T_2$ TERMS IN AXIALLY SYMMETRIC CRYSTAL FIELDS [5]

As for the $^3T_1$ terms the calculations for the $^5T_2$ terms will merely be outlined and some of the results discussed. The calculation outlined assumes that all contributions from excited crystal field terms can be ignored. The wave-functions for the $^5T_2$ term appropriate to tetragonal symmetry (the corresponding wave-functions for trigonal symmetry are shown in Chapter 2) are, as $|M_L, M_S\rangle$:

$$\psi_1 = |1, 2\rangle \qquad\qquad \psi_9 = |-1, -1\rangle$$
$$\psi_2 = \sqrt{\tfrac{1}{2}}(|2, 1\rangle - |-2, 1\rangle) \qquad \psi_{10} = |1, -1\rangle$$
$$\psi_3 = |-1, 0\rangle \qquad\qquad \psi_{11} = \sqrt{\tfrac{1}{2}}(|2, -2\rangle - |-2, -2\rangle)$$
$$\psi_4 = |-1, -2\rangle \qquad\qquad \psi_{12} = |-1, 1\rangle$$
$$\psi_5 = \sqrt{\tfrac{1}{2}}(|2, -1\rangle - |-2 - 1\rangle) \qquad \psi_{13} = \sqrt{\tfrac{1}{2}}(|2, 2\rangle - |-2, 2\rangle)$$
$$\psi_6 = |1, 0\rangle \qquad\qquad \psi_{14} = |1, -2\rangle$$
$$\psi_7 = |1, 1\rangle \qquad\qquad \psi_{15} = |-1, 2\rangle$$
$$\psi_8 = \sqrt{\tfrac{1}{2}}(|2, 0\rangle - |-2, 0\rangle)$$

Using these fifteen wave-functions and the perturbing operator $V_{\text{tetrag}} + \lambda \hat{L} \cdot \hat{S}$ gives a secular determinant which factorizes into the following sub-determinants

$$
\begin{array}{c|ccc}
 & |\psi_1\rangle & |\psi_2\rangle & |\psi_3\rangle \\
\hline
\langle\psi_1| & \dfrac{\Delta}{3}+2\lambda-E & \sqrt{2}\lambda & 0 \\
\langle\psi_2| & \sqrt{2}\lambda & -\dfrac{2\Delta}{3}-E & -\sqrt{3}\lambda \\
\langle\psi_3| & 0 & -\sqrt{3}\lambda & \dfrac{\Delta}{3}-E
\end{array} = 0
$$

$$
\begin{array}{c|ccc}
 & |\psi_4\rangle & |\psi_5\rangle & |\psi_6\rangle \\
\hline
\langle\psi_4| & \dfrac{\Delta}{3}+2\lambda-E & -\sqrt{2}\lambda & 0 \\
\langle\psi_5| & -\sqrt{2}\lambda & -\dfrac{2\Delta}{3}-E & \sqrt{3}\lambda \\
\langle\psi_6| & 0 & \sqrt{3}\lambda & \dfrac{\Delta}{3}-E
\end{array} = 0
$$

$$
\begin{array}{c|ccc}
 & |\psi_7\rangle & |\psi_8\rangle & |\psi_9\rangle \\
\hline
\langle\psi_7| & \dfrac{\Delta}{3}+\lambda-E & \sqrt{3}\lambda & 0 \\
\langle\psi_8| & \sqrt{3}\lambda & -\dfrac{2\Delta}{3}-E & -\sqrt{3}\lambda \\
\langle\psi_9| & 0 & -\sqrt{3}\lambda & \dfrac{\Delta}{3}+\lambda-E
\end{array} = 0 \qquad (5.23)
$$

$$
\begin{array}{c|cc}
 & |\psi_{10}\rangle & |\psi_{11}\rangle \\
\hline
\langle\psi_{10}| & \dfrac{\Delta}{3}-\lambda-E & \sqrt{2}\lambda \\
\langle\psi_{11}| & \sqrt{2}\lambda & -\dfrac{2\Delta}{3}-E
\end{array} = 0
$$

$$
\begin{array}{c|cc}
 & |\psi_{12}\rangle & |\psi_{13}\rangle \\
\hline
\langle\psi_{12}| & \dfrac{\Delta}{3}-\lambda-E & -\sqrt{2}\lambda \\
\langle\psi_{13}| & -\sqrt{2}\lambda & -\dfrac{2\Delta}{3}-E
\end{array} = 0
$$

$$
\begin{array}{c|c}
 & |\psi_{14}\rangle \\
\hline
\langle\psi_{14}| & \dfrac{\Delta}{3}-2\lambda-E
\end{array} = 0
$$

$$
\begin{array}{c|c}
 & |\psi_{15}\rangle \\
\hline
\langle\psi_{15}| & \dfrac{\Delta}{3}-2\lambda-E
\end{array} = 0
$$

The energy levels arising from these determinants cannot all be given in a closed form since the cubic equations involved do not factorize. The usual procedure in such cases is to solve the determinant numerically on a computer for the desired ranges of the parameters. The energies and wave-functions obtained are then used to calculate the magnetic susceptibilities as outlined in Section 5.2.1. Values of $\bar{\mu}$ for certain ranges of the parameters v, $\mathbf{k}$ and kT/λ are shown in Table 5.4, and the variation of $\bar{\mu}$ against kT/λ for some particular values of v and $\mathbf{k}$ are illustrated in Figs. 5.9–10.

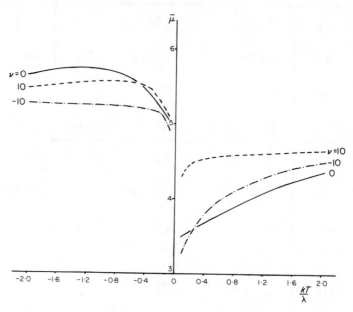

Figure 5.9. The variation of $\bar{\mu}$ vs. kT/λ for the $^5T_2$ term for various values of v, with $\mathbf{k} = 1\cdot0$.

(a) *The variation of $\bar{\mu}$ with v* (illustrated in Fig. 5.9). When kT/λ is positive $\bar{\mu}$ tends towards the spin-only value of 4·90 B.M. at the higher values of kT/λ as v increases in magnitude. Compared with the undistorted situation there is less variation of $\bar{\mu}$ with temperature when v is positive but a slightly greater variation when v is negative. With kT/λ negative the effect on $\bar{\mu}$ of changing v is very similar to that discussed above, the only change being that the temperature variation for v both positive and negative is now less than for $v = 0$.

(b) *The variation of $\bar{\mu}$ with $\mathbf{k}$* (illustrated in Fig. 5.10). Examination of Table 5.4 shows that at any particular value of v two distinct trends are discernable. On the one hand when kT/λ is positive $\bar{\mu}$ tends towards the spin-only value as $\mathbf{k}$ is reduced from unity for all values of kT/λ. In contrast to this, when kT/λ is

Table 5.4. The values of $\bar{\mu}$ for the $^5T_2$ term as a function of kT/λ, v, and **k**. (Reproduced with permission from reference 5.)

Calculated magnetic moments.

k = 1·0

kT/λ	0·1	0·2	0·3	0·5	0·75	1·0	1·5	2·0	3·0	−0·1	−0·2	−0·3	−0·5	−0·75	−1·0	−1·5	−2·0	−3·0
v 10	4·32	4·47	4·52	4·56	4·59	4·60	4·63	4·66	4·71	5·12	5·30	5·42	5·53	5·57	5·56	5·52	5·47	5·41
5	3·94	4·14	4·21	4·28	4·33	4·37	4·43	4·50	4·63	5·06	5·28	5·40	5·53	5·60	5·62	5·60	5·56	5·48
3	3·73	3·91	3·98	4·06	4·13	4·19	4·31	4·43	4·60	5·02	5·25	5·37	5·54	5·63	5·66	5·65	5·60	5·50
2	3·62	3·77	3·84	3·92	4·00	4·09	4·26	4·40	4·59	5·02	5·23	5·36	5·54	5·65	5·69	5·68	5·62	5·52
1	3·54	3·64	3·70	3·79	3·91	4·02	4·22	4·38	4·59	5·05	5·22	5·35	5·54	5·67	5·71	5·70	5·64	5·52
0	3·52	3·58	3·63	3·74	3·87	4·00	4·21	4·38	4·58	5·08	5·22	5·34	5·54	5·67	5·72	5·70	5·64	5·53
−1	3·44	3·59	3·67	3·77	3·90	4·02	4·22	4·38	4·59	5·04	5·22	5·35	5·54	5·66	5·71	5·69	5·64	5·52
−2	3·38	3·59	3·69	3·82	3·94	4·05	4·24	4·39	4·59	4·95	5·22	5·35	5·52	5·63	5·67	5·66	5·62	5·51
−3	3·34	3·57	3·70	3·85	3·98	4·09	4·27	4·41	4·60	4·90	5·22	5·35	5·49	5·57	5·61	5·61	5·58	5·50
−5	3·31	3·55	3·70	3·89	4·04	4·15	4·32	4·44	4·61	4·89	5·21	5·32	5·41	5·44	5·47	5·49	5·49	5·45
−10	3·28	3·52	3·69	3·91	4·08	4·21	4·39	4·50	4·65	4·95	5·14	5·19	5·22	5·23	5·24	5·26	5·27	5·29

k = 0·9

kT/λ	0·1	0·2	0·3	0·5	0·75	1·0	1·5	2·0	3·0	−0·1	−0·2	−0·3	−0·5	−0·75	−1·0	−1·5	−2·0	−3·0
v 10	4·35	4·50	4·55	4·59	4·61	4·63	4·65	4·67	4·71	5·03	5·21	5·32	5·44	5·48	5·48	5·44	5·40	5·34
5	4·00	4·19	4·27	4·33	4·38	4·41	4·46	4·53	4·64	4·97	5·18	5·30	5·44	5·51	5·53	5·51	5·48	5·40
3	3·80	3·98	4·05	4·12	4·18	4·24	4·35	4·45	4·61	4·93	5·16	5·28	5·44	5·54	5·66	5·65	5·60	5·50
2	3·71	3·86	3·92	3·99	4·07	4·15	4·30	4·43	4·60	4·94	5·15	5·27	5·44	5·55	5·59	5·58	5·54	5·44
1	3·64	3·74	3·79	3·88	3·98	4·09	4·27	4·41	4·60	4·97	5·14	5·26	5·44	5·56	5·61	5·60	5·55	5·45
0	3·63	3·68	3·73	3·83	3·95	4·07	4·26	4·41	4·59	5·01	5·13	5·25	5·44	5·57	5·62	5·61	5·55	5·45
−1	3·55	3·69	3·76	3·86	3·97	4·08	4·27	4·41	4·60	4·96	5·13	5·26	5·44	5·56	5·61	5·60	5·55	5·45
−2	3·47	3·68	3·79	3·90	4·01	4·11	4·29	4·42	4·60	4·87	5·14	5·27	5·43	5·53	5·57	5·57	5·53	5·44
−3	3·44	3·66	3·79	3·93	4·05	4·15	4·31	4·44	4·61	4·82	5·14	5·27	5·41	5·48	5·52	5·52	5·50	5·42
−5	3·40	3·64	3·79	3·96	4·10	4·20	4·36	4·47	4·62	4·83	5·14	5·25	5·34	5·38	5·40	5·42	5·41	5·38
−10	3·38	3·61	3·77	3·98	4·14	4·26	4·42	4·53	4·66	4·91	5·10	5·15	5·18	5·19	5·20	5·21	5·23	5·24

k = 0·8

kT/λ	0·1	0·2	0·3	0·5	0·75	1·0	1·5	2·0	3·0	−0·1	−0·2	−0·3	−0·5	−0·75	−1·0	−1·5	−2·0	−3·0
v 10	4·39	4·53	4·58	4·62	4·64	4·65	4·67	4·69	4·72	4·93	5·12	5·23	5·35	5·39	5·40	5·37	5·34	5·28
5	4·06	4·25	4·32	4·38	4·42	4·45	4·50	4·55	4·65	4·88	5·09	5·21	5·35	5·42	5·44	5·43	5·40	5·34
3	3·88	4·05	4·12	4·19	4·25	4·29	4·39	4·49	4·62	4·85	5·07	5·19	5·35	5·44	5·47	5·47	5·43	5·36
2	3·79	3·94	4·00	4·07	4·14	4·21	4·35	4·46	4·62	4·86	5·06	5·18	5·34	5·45	5·49	5·49	5·45	5·37
1	3·74	3·84	3·89	3·96	4·06	4·15	4·32	4·45	4·61	4·90	5·05	5·17	5·34	5·46	5·51	5·50	5·46	5·37
0	3·74	3·79	3·83	3·92	4·03	4·13	4·31	4·44	4·61	4·93	5·05	5·16	5·34	5·47	5·51	5·51	5·47	5·37
−1	3·65	3·79	3·86	3·95	4·05	4·15	4·32	4·44	4·61	4·88	5·05	5·17	5·34	5·46	5·50	5·50	5·46	5·37
−2	3·57	3·78	3·88	3·99	4·09	4·18	4·34	4·46	4·61	4·80	5·06	5·18	5·34	5·43	5·47	5·48	5·44	5·36
−3	3·53	3·76	3·88	4·01	4·12	4·21	4·36	4·47	4·62	4·75	5·06	5·19	5·32	5·40	5·43	5·44	5·42	5·35
−5	3·50	3·73	3·87	4·04	4·17	4·26	4·40	4·50	4·63	4·77	5·07	5·18	5·27	5·31	5·33	5·35	5·34	5·31
−10	3·47	3·70	3·86	4·06	4·21	4·32	4·46	4·55	4·67	4·87	5·06	5·11	5·14	5·15	5·16	5·17	5·18	5·20

k = 0·7

kT/λ	0·1	0·2	0·3	0·5	0·75	1·0	1·5	2·0	3·0	−0·1	−0·2	−0·3	−0·5	−0·75	−1·0	−1·5	−2·0	−3·0
v 10	4·42	4·57	4·61	4·65	4·66	4·68	4·69	4·71	4·74	4·84	5·02	5·14	5·26	5·31	5·32	5·30	5·27	5·22
5	4·12	4·31	4·38	4·43	4·47	4·49	4·54	4·58	4·67	4·79	5·00	5·12	5·26	5·33	5·35	5·35	5·32	5·27
3	3·96	4·13	4·20	4·26	4·31	4·35	4·44	4·52	4·64	4·76	4·99	5·10	5·25	5·34	5·38	5·38	5·35	5·29
2	3·88	4·03	4·09	4·15	4·21	4·28	4·40	4·50	4·64	4·77	4·97	5·09	5·25	5·35	5·39	5·40	5·37	5·29
1	3·84	3·94	3·98	4·05	4·14	4·23	4·37	4·48	4·63	4·82	4·97	5·08	5·24	5·36	5·41	5·41	5·38	5·30
0	3·85	3·90	3·94	4·02	4·12	4·21	4·36	4·48	4·63	4·85	4·96	5·07	5·24	5·36	5·41	5·42	5·38	5·30
−1	3·75	3·89	3·96	4·04	4·13	4·22	4·37	4·48	4·63	4·81	4·97	5·08	5·24	5·36	5·41	5·41	5·37	5·30
−2	3·67	3·87	3·97	4·07	4·17	4·25	4·39	4·49	4·63	4·72	4·97	5·09	5·24	5·34	5·38	5·39	5·36	5·29
−3	3·63	3·85	3·97	4·10	4·20	4·28	4·41	4·51	4·64	4·68	4·99	5·11	5·24	5·31	5·34	5·36	5·34	5·28
−5	3·59	3·82	3·96	4·12	4·24	4·32	4·45	4·53	4·65	4·71	5·01	5·12	5·20	5·24	5·26	5·28	5·28	5·25
−10	3·56	3·79	3·94	4·13	4·27	4·37	4·50	4·58	4·68	4·83	5·02	5·07	5·10	5·11	5·12	5·13	5·14	5·15

k = 0·6

kT/λ	0·1	0·2	0·3	0·5	0·75	1·0	1·5	2·0	3·0	−0·1	−0·2	−0·3	−0·5	−0·75	−1·0	−1·5	−2·0	−3·0
v 10	4·46	4·60	4·64	4·68	4·69	4·70	4·71	4·73	4·75	4·74	4·93	5·05	5·17	5·23	5·24	5·23	5·20	5·17
5	4·18	4·36	4·43	4·48	4·52	4·54	4·57	4·61	4·69	4·70	4·91	5·03	5·17	5·24	5·27	5·27	5·25	5·20
3	4·03	4·20	4·27	4·33	4·37	4·41	4·49	4·56	4·67	4·68	4·90	5·01	5·16	5·25	5·29	5·29	5·27	5·22
2	3·97	4·12	4·17	4·23	4·29	4·35	4·45	4·54	4·66	4·69	4·89	5·00	5·15	5·25	5·30	5·31	5·28	5·22
1	3·94	4·03	4·08	4·15	4·22	4·30	4·43	4·53	4·65	4·74	4·88	4·99	5·15	5·26	5·31	5·32	5·29	5·23
0	3·96	4·00	4·04	4·11	4·20	4·28	4·42	4·52	4·65	4·77	4·88	4·98	5·14	5·26	5·31	5·32	5·30	5·23
−1	3·85	4·00	4·06	4·13	4·22	4·29	4·43	4·53	4·65	4·73	4·88	4·99	5·15	5·26	5·31	5·32	5·29	5·23
−2	3·77	3·97	4·06	4·16	4·25	4·32	4·44	4·53	4·66	4·65	4·89	5·00	5·15	5·25	5·29	5·30	5·28	5·22
−3	3·73	3·94	4·06	4·18	4·27	4·35	4·46	4·55	4·66	4·61	4·91	5·02	5·15	5·22	5·26	5·27	5·26	5·21
−5	3·69	3·91	4·05	4·20	4·31	4·38	4·49	4·57	4·67	4·64	4·94	5·05	5·13	5·17	5·19	5·21	5·21	5·19
−10	3·66	3·88	4·03	4·21	4·34	4·43	4·54	4·62	4·70	4·79	4·98	5·03	5·07	5·08	5·08	5·09	5·10	5·11

k = 0·5

kT/λ	0·1	0·2	0·3	0·5	0·75	1·0	1·5	2·0	3·0	−0·1	−0·2	−0·3	−0·5	−0·75	−1·0	−1·5	−2·0	−3·0
v 10	4·50	4·64	4·68	4·71	4·72	4·73	4·74	4·75	4·77	4·65	4·84	4·96	5·08	5·14	5·16	5·16	5·14	5·11
5	4·24	4·42	4·49	4·54	4·56	4·58	4·62	4·65	4·71	4·61	4·82	4·94	5·07	5·15	5·18	5·19	5·18	5·14
3	4·11	4·28	4·34	4·40	4·44	4·47	4·54	4·60	4·69	4·59	4·81	4·92	5·06	5·15	5·19	5·21	5·20	5·15
2	4·06	4·20	4·26	4·31	4·37	4·42	4·51	4·58	4·68	4·61	4·80	4·91	5·06	5·16	5·20	5·22	5·20	5·16
1	4·04	4·13	4·18	4·24	4·31	4·37	4·49	4·57	4·68	4·66	4·80	4·90	5·05	5·16	5·21	5·23	5·21	5·16
0	4·08	4·11	4·14	4·21	4·29	4·36	4·48	4·57	4·68	4·70	4·80	4·89	5·05	5·16	5·21	5·23	5·21	5·16
−1	3·95	4·10	4·16	4·23	4·30	4·37	4·49	4·57	4·68	4·65	4·80	4·90	5·05	5·16	5·21	5·23	5·21	5·16
−2	3·87	4·07	4·16	4·25	4·33	4·39	4·50	4·58	4·68	4·57	4·81	4·92	5·06	5·15	5·19	5·21	5·20	5·16
−3	3·82	4·04	4·15	4·27	4·35	4·42	4·52	4·59	4·69	4·53	4·83	4·94	5·07	5·14	5·17	5·19	5·19	5·15
−5	3·79	4·00	4·14	4·28	4·38	4·45	4·55	4·61	4·70	4·58	4·88	4·98	5·07	5·11	5·13	5·14	5·15	5·13
−10	3·75	3·97	4·11	4·29	4·41	4·49	4·59	4·65	4·72	4·75	4·94	5·00	5·03	5·04	5·04	5·05	5·06	5·06

negative $\bar{\mu}$ at all values of kT/λ is simply reduced as **k** is reduced. These trends are summarized in Fig. 5.10 for the particular case of $v = 10$.

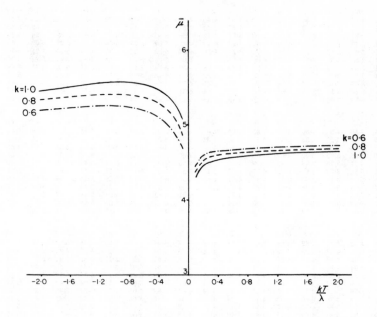

Figure 5.10. The variation of $\bar{\mu}$ vs. kT/λ for the $^5T_2$ term for various values of **k**, with $v = 10$.

5.2.4. THE MAGNETIC PROPERTIES OF $^4T_1$ TERMS IN AXIALLY SYMMETRIC CRYSTAL FIELDS [6]

When considering a tetragonal distortion the orbital wave-functions used to specify the $^4T_1$ term are the same as those used for the $^3T_1$ term. Each orbital part has now to be taken with the spin functions $|\pm\tfrac{3}{2}\rangle$, $|\pm\tfrac{1}{2}\rangle$ giving:

$$\psi_1 = |A, \tfrac{3}{2}\rangle, \qquad \psi_2 = |0, \tfrac{1}{2}\rangle, \qquad \psi_3 = |-A, -\tfrac{1}{2}\rangle$$
$$\psi_4 = |-A, -\tfrac{3}{2}\rangle, \qquad \psi_5 = |0, -\tfrac{1}{2}\rangle, \qquad \psi_6 = |A, \tfrac{1}{2}\rangle$$
$$\psi_7 = |A, -\tfrac{1}{2}\rangle, \qquad \psi_8 = |0, -\tfrac{3}{2}\rangle, \qquad \psi_9 = |-A, \tfrac{1}{2}\rangle$$
$$\psi_{10} = |0, \tfrac{3}{2}\rangle, \qquad \psi_{11} = |A, -\tfrac{3}{2}\rangle, \quad \psi_{12} = |-A, \tfrac{3}{2}\rangle.$$

Using the restrictions given in 5.2.2, perturbation of the $^4T_1$ term by an axially symmetric crystal field and spin–orbit coupling leads to a secular determinant which factorizes into the following sub-determinants:

$$
\left.
\begin{array}{c|ccc}
 & |\pm A, \pm\tfrac{3}{2}\rangle & |0, \pm\tfrac{1}{2}\rangle & |\mp A, \mp\tfrac{1}{2}\rangle \\
\hline
\langle\pm A, \pm\tfrac{3}{2}| & \dfrac{\Delta}{3}+\dfrac{3A\lambda}{2}-E & \sqrt{\tfrac{3}{2}}A\lambda & 0 \\
\langle 0, \pm\tfrac{1}{2}| & \sqrt{\tfrac{3}{2}}A\lambda & -\dfrac{2\Delta}{3}-E & \sqrt{2}A\lambda \\
\langle\mp A, \mp\tfrac{1}{2}| & 0 & \sqrt{2}A\lambda & \dfrac{\Delta}{3}+\dfrac{A\lambda}{2}-E \\
\end{array}
\right| = 0
$$

$$
\left.
\begin{array}{c|cc}
 & |\pm A, \mp\tfrac{1}{2}\rangle & |0, \mp\tfrac{3}{2}\rangle \\
\hline
\langle\pm A, \mp\tfrac{1}{2}| & \dfrac{\Delta}{3}+\dfrac{A\lambda}{2}-E & \sqrt{\tfrac{3}{2}}A\lambda \\
\langle 0, \mp\tfrac{3}{2}| & \sqrt{\tfrac{3}{2}}A\lambda & -\dfrac{2\Delta}{3}-E \\
\end{array}
\right| = 0
$$

$$
\left.
\begin{array}{c|c}
 & |\pm A, \mp\tfrac{3}{2}\rangle \\
\hline
\langle\pm A, \mp\tfrac{3}{2}| & \dfrac{\Delta}{3}-\dfrac{3A\lambda}{2}-E \\
\end{array}
\right| = 0
\tag{5.24}
$$

The solution of these determinants and the subsequent calculation of the magnetic susceptibilities follows the method already outlined. Tables for the variation of $\bar\mu$ with the various parameters are available in the literature [6] and are summarized in Table 5.5. Some examples of the variation of $\bar\mu$ with the various parameters are illustrated in Figs. 5.11–5.13.

(a) *The variation of $\bar\mu$ with v* (illustrated in Fig. 5.11). At the higher values of $|kT/\lambda|$ increasing $|v|$ makes $\bar\mu$ tend towards the spin-only value of 3·87 B.M. Thus when kT/λ is positive this corresponds to an increase in $\bar\mu$ with increasing distortion whilst the opposite trend occurs for kT/λ negative. The temperature variation of $\bar\mu$ depends on the sign of both kT/λ and v. When kT/λ and v are positive $\bar\mu$ is always greater than that for the undistorted situation and has the result of causing less temperature variation in $\bar\mu$. However, when kT/λ is positive and v is negative the details depend upon the particular values not only of v but also of A and $\mathbf{k}$, although the general tendency is for there to be less temperature variation in $\bar\mu$.

(b) *The variation of $\bar\mu$ with $\mathbf{k}$* (illustrated in Fig. 5.12). The effect of reducing $\mathbf{k}$ from unity follows the same trends outlined for the $^5T_2$ term. When kT/λ is positive, $\bar\mu$ increases towards the spin-only value for all values of v, A and kT/λ.

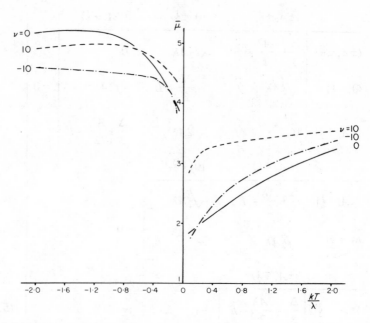

Figure 5.11. The variation of $\bar{\mu}$ vs. kT/λ for the $^4T_1$ term for various values of v, with $\mathbf{k} = 1\cdot0$ and $A = 1\cdot5$.

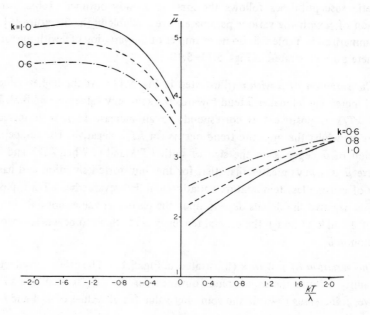

Figure 5.12. The variation of $\bar{\mu}$ vs. kT/λ for the $^4T_1$ term for various values of $\mathbf{k}$ with $v = 0$ and $A = 1\cdot5$.

Table 5.5. *The values of $\bar{\mu}$ for the $^4T_1$ term as a function of kT/λ (values at the top of the columns), v, $\mathbf{k}$, and A.*
(Reproduced with permission from reference 6.)

Calculated magnetic moments (B.M.)

$A = 1.5,\ \mathbf{k} = 1.0$

v	2.0	1.7	1.3	1.0	0.8	0.6	0.4	0.3	0.2	0.1	−2.0	−1.7	−1.3	−1.0	−0.8	−0.6	−0.4	−0.3	−0.2	−0.1
10	3.55	3.51	3.46	3.42	3.38	3.34	3.28	3.24	3.14	2.93	4.92	4.95	4.98	4.98	4.96	4.90	4.78	4.68	4.55	4.41
5	3.37	3.29	3.18	3.10	3.04	2.97	2.88	2.80	2.69	2.47	5.06	5.07	5.08	5.06	5.00	4.90	4.70	4.56	4.40	4.21
3	3.29	3.18	3.01	2.88	2.79	2.69	2.59	2.51	2.41	2.21	5.12	5.14	5.15	5.10	5.03	4.89	4.66	4.49	4.30	4.09
2	3.26	3.13	2.93	2.76	2.65	2.53	2.41	2.34	2.24	2.07	5.15	5.18	5.18	5.13	5.05	4.89	4.63	4.45	4.25	4.04
1	3.24	3.10	2.88	2.68	2.54	2.40	2.24	2.16	2.08	1.95	5.17	5.20	5.21	5.15	5.06	4.89	4.61	4.43	4.22	4.00
0	3.24	3.09	2.86	2.66	2.51	2.35	2.17	2.08	1.98	1.88	5.18	5.21	5.22	5.16	5.07	4.89	4.61	4.42	4.21	4.00
−1	3.24	3.10	2.88	2.68	2.53	2.38	2.21	2.12	2.02	1.85	5.17	5.20	5.20	5.15	5.06	4.89	4.61	4.43	4.22	4.00
−2	3.25	3.12	2.91	2.72	2.58	2.43	2.27	2.16	2.03	1.81	5.14	5.16	5.16	5.11	5.03	4.88	4.62	4.45	4.24	4.02
−3	3.27	3.14	2.94	2.76	2.63	2.48	2.30	2.18	2.02	1.79	5.09	5.10	5.09	5.04	4.97	4.85	4.63	4.47	4.26	4.03
−5	3.31	3.19	3.00	2.83	2.69	2.53	2.32	2.18	2.00	1.77	4.95	4.94	4.91	4.86	4.82	4.75	4.61	4.48	4.28	4.03
−10	3.37	3.26	3.08	2.90	2.75	2.56	2.32	2.17	1.98	1.76	4.59	4.57	4.53	4.51	4.49	4.47	4.42	4.36	4.24	3.96

$A = 1.5,\ \mathbf{k} = 0.8$

v	2.0	1.7	1.3	1.0	0.8	0.6	0.4	0.3	0.2	0.1	−2.0	−1.7	−1.3	−1.0	−0.8	−0.6	−0.4	−0.3	−0.2	−0.1
10	3.57	3.54	3.51	3.48	3.45	3.42	3.38	3.33	3.25	3.04	4.67	4.69	4.71	4.71	4.68	4.62	4.50	4.40	4.28	4.14
5	3.39	3.33	3.25	3.19	3.15	3.10	3.02	2.96	2.86	2.64	4.77	4.78	4.79	4.76	4.71	4.61	4.43	4.30	4.14	3.98
3	3.31	3.22	3.09	2.99	2.92	2.85	2.77	2.71	2.62	2.43	4.82	4.84	4.84	4.79	4.73	4.60	4.38	4.23	4.06	3.88
2	3.28	3.18	3.02	2.89	2.80	2.71	2.62	2.56	2.48	2.32	4.84	4.86	4.86	4.81	4.74	4.60	4.36	4.20	4.02	3.83
1	3.26	3.15	2.97	2.82	2.71	2.59	2.48	2.42	2.35	2.24	4.86	4.88	4.88	4.83	4.74	4.59	4.34	4.17	3.99	3.79
0	3.25	3.14	2.95	2.79	2.68	2.55	2.42	2.35	2.28	2.21	4.87	4.89	4.89	4.83	4.75	4.59	4.33	4.17	3.98	3.79
−1	3.26	3.15	2.97	2.81	2.70	2.58	2.45	2.38	2.30	2.14	4.86	4.88	4.88	4.83	4.74	4.59	4.34	4.17	3.99	3.81
−2	3.27	3.17	3.00	2.85	2.74	2.63	2.50	2.41	2.29	2.09	4.83	4.85	4.84	4.80	4.72	4.59	4.36	4.20	4.01	3.83
−3	3.29	3.19	3.03	2.89	2.78	2.67	2.52	2.42	2.27	2.06	4.79	4.80	4.79	4.75	4.68	4.57	4.38	4.23	4.04	3.84
−5	3.33	3.23	3.09	2.95	2.84	2.71	2.53	2.41	2.25	2.04	4.68	4.68	4.65	4.62	4.58	4.51	4.38	4.26	4.08	3.84
−10	3.39	3.31	3.16	3.01	2.89	2.74	2.53	2.40	2.23	2.03	4.41	4.40	4.37	4.35	4.34	4.32	4.28	4.22	4.10	3.83

$A = 1.5,\ \mathbf{k} = 0.6$

v	2.0	1.7	1.3	1.0	0.8	0.6	0.4	0.3	0.2	0.1	−2.0	−1.7	−1.3	−1.0	−0.8	−0.6	−0.4	−0.3	−0.2	−0.1
10	3.61	3.59	3.57	3.55	3.53	3.51	3.48	3.44	3.36	3.16	4.42	4.44	4.45	4.44	4.41	4.35	4.22	4.13	4.01	3.88
5	3.45	3.41	3.35	3.31	3.28	3.24	3.18	3.13	3.04	2.82	4.50	4.50	4.50	4.47	4.42	4.33	4.16	4.04	3.90	3.74
3	3.37	3.31	3.21	3.14	3.09	3.04	2.98	2.93	2.84	2.65	4.53	4.54	4.53	4.49	4.42	4.31	4.11	3.98	3.82	3.66
2	3.35	3.27	3.15	3.05	2.99	2.92	2.85	2.81	2.74	2.57	4.55	4.56	4.55	4.50	4.43	4.30	4.09	3.95	3.79	3.62
1	3.33	3.24	3.11	2.99	2.91	2.83	2.74	2.69	2.64	2.53	4.56	4.57	4.56	4.51	4.43	4.29	4.07	3.92	3.76	3.59
0	3.32	3.24	3.10	2.97	2.88	2.79	2.69	2.64	2.59	2.54	4.56	4.58	4.57	4.51	4.43	4.29	4.06	3.92	3.75	3.58
−1	3.33	3.24	3.11	2.99	2.90	2.81	2.72	2.66	2.59	2.44	4.56	4.57	4.56	4.51	4.43	4.29	4.07	3.92	3.76	3.59
−2	3.34	3.26	3.13	3.02	2.94	2.85	2.75	2.68	2.57	2.37	4.54	4.55	4.54	4.49	4.42	4.30	4.09	3.95	3.79	3.61
−3	3.36	3.28	3.16	3.05	2.97	2.88	2.76	2.67	2.54	2.35	4.51	4.52	4.50	4.46	4.40	4.30	4.12	3.98	3.82	3.63
−5	3.39	3.32	3.21	3.10	3.02	2.91	2.77	2.66	2.51	2.32	4.44	4.43	4.41	4.38	4.34	4.28	4.16	4.04	3.88	3.66
−10	3.45	3.38	3.27	3.16	3.06	2.93	2.76	2.64	2.48	2.30	4.25	4.24	4.22	4.21	4.20	4.18	4.13	4.08	3.96	3.70

$A = 1.4,\ \mathbf{k} = 1.0$

v	2.0	1.7	1.3	1.0	0.8	0.6	0.4	0.3	0.2	0.1	−2.0	−1.7	−1.3	−1.0	−0.8	−0.6	−0.4	−0.3	−0.2	−0.1
10	3.59	3.56	3.51	3.47	3.44	3.41	3.36	3.32	3.24	3.03	4.81	4.84	4.87	4.88	4.87	4.82	4.71	4.62	4.49	4.35
5	3.42	3.35	3.25	3.18	3.12	3.06	2.98	2.91	2.81	2.59	4.94	4.96	4.97	4.96	4.92	4.83	4.65	4.52	4.35	4.16
3	3.35	3.24	3.09	2.96	2.88	2.79	2.70	2.63	2.53	2.33	5.00	5.03	5.04	5.01	4.95	4.83	4.61	4.45	4.26	4.05
2	3.32	3.20	3.01	2.85	2.74	2.63	2.52	2.45	2.36	2.19	5.03	5.06	5.07	5.04	4.97	4.83	4.59	4.41	4.21	3.99
1	3.30	3.17	2.96	2.77	2.64	2.49	2.35	2.27	2.19	2.07	5.05	5.08	5.10	5.06	4.99	4.84	4.57	4.38	4.17	3.95
0	3.29	3.16	2.94	2.75	2.60	2.44	2.27	2.18	2.09	2.00	5.06	5.09	5.11	5.07	4.99	4.84	4.56	4.37	4.16	3.94
−1	3.30	3.17	2.95	2.76	2.62	2.48	2.32	2.23	2.13	1.96	5.05	5.08	5.10	5.06	4.98	4.83	4.57	4.38	4.17	3.95
−2	3.31	3.19	2.98	2.81	2.67	2.53	2.37	2.27	2.13	1.91	5.02	5.04	5.05	5.01	4.95	4.82	4.58	4.41	4.20	3.97
−3	3.33	3.21	3.02	2.85	2.72	2.58	2.40	2.28	2.12	1.89	4.97	4.98	4.98	4.94	4.88	4.78	4.59	4.43	4.22	3.98
−5	3.36	3.25	3.08	2.91	2.78	2.62	2.42	2.29	2.11	1.88	4.83	4.83	4.80	4.76	4.73	4.67	4.55	4.44	4.25	3.98
−10	3.42	3.32	3.15	2.98	2.84	2.66	2.43	2.28	2.09	1.86	4.50	4.48	4.45	4.43	4.41	4.39	4.36	4.31	4.21	3.94

$A = 1.4,\ \mathbf{k} = 0.8$

v	2.0	1.7	1.3	1.0	0.8	0.6	0.4	0.3	0.2	0.1	−2.0	−1.7	−1.3	−1.0	−0.8	−0.6	−0.4	−0.3	−0.2	−0.1
10	3.60	3.58	3.55	3.52	3.50	3.48	3.44	3.40	3.33	3.14	4.58	4.60	4.63	4.63	4.62	4.57	4.45	4.36	4.24	4.10
5	3.44	3.39	3.32	3.26	3.22	3.18	3.11	3.06	2.96	2.75	4.68	4.70	4.70	4.69	4.65	4.56	4.39	4.27	4.11	3.94
3	3.37	3.28	3.16	3.07	3.01	2.94	2.87	2.81	2.72	2.53	4.73	4.75	4.75	4.72	4.67	4.56	4.35	4.20	4.03	3.84
2	3.34	3.24	3.09	2.97	2.89	2.80	2.71	2.66	2.58	2.42	4.75	4.77	4.78	4.74	4.68	4.55	4.33	4.17	3.99	3.79
1	3.32	3.21	3.05	2.90	2.79	2.68	2.57	2.51	2.44	2.33	4.76	4.79	4.80	4.76	4.69	4.55	4.31	4.15	3.96	3.76
0	3.31	3.21	3.03	2.87	2.76	2.64	2.51	2.44	2.37	2.30	4.77	4.80	4.81	4.77	4.69	4.55	4.30	4.13	3.95	3.75
−1	3.32	3.21	3.04	2.89	2.78	2.67	2.54	2.47	2.39	2.23	4.76	4.79	4.80	4.76	4.69	4.55	4.31	4.15	3.96	3.76
−2	3.33	3.23	3.07	2.93	2.83	2.71	2.58	2.50	2.38	2.18	4.74	4.76	4.76	4.72	4.66	4.54	4.33	4.17	3.98	3.78
−3	3.35	3.25	3.10	2.97	2.87	2.75	2.61	2.51	2.36	2.15	4.70	4.71	4.71	4.67	4.62	4.52	4.33	4.20	4.01	3.80
−5	3.38	3.30	3.16	3.03	2.92	2.79	2.62	2.50	2.33	2.13	4.60	4.59	4.57	4.54	4.51	4.46	4.34	4.24	4.06	3.81
−10	3.44	3.36	3.23	3.09	2.97	2.82	2.62	2.49	2.32	2.11	4.35	4.33	4.31	4.29	4.28	4.27	4.23	4.19	4.09	3.83

Table 5.5 (continued)

$A = 1\cdot4$, $\mathbf{k} = 0\cdot6$

v	2·0	1·7	1·3	1·0	0·8	0·6	0·4	0·3	0·2	0·1	−2·0	−1·7	−1·3	−1·0	−0·8	−0·6	−0·4	−0·3	−0·2	−0·1
10	3·64	3·62	3·61	3·59	3·57	3·56	3·53	3·50	3·43	3·24	4·37	4·38	4·39	4·39	4·36	4·31	4·19	4·10	3·98	3·85
5	3·49	3·46	3·40	3·37	3·34	3·31	3·26	3·21	3·12	2·91	4·43	4·44	4·44	4·42	4·38	4·30	4·14	4·02	3·88	3·72
3	3·43	3·36	3·28	3·21	3·16	3·11	3·05	3·01	2·93	2·73	4·47	4·48	4·48	4·44	4·29	4·28	4·10	3·96	3·81	3·64
2	3·40	3·33	3·22	3·12	3·06	3·00	2·93	2·89	2·82	2·65	4·48	4·49	4·49	4·45	4·39	4·28	4·07	3·93	3·77	3·60
1	3·38	3·30	3·18	3·06	2·98	2·90	2·81	2·77	2·71	2·60	4·49	4·51	4·51	4·46	4·40	4·27	4·05	3·91	3·74	3·57
0	3·38	3·30	3·16	3·04	2·96	2·86	2·76	2·71	2·66	2·61	4·50	4·51	4·51	4·47	4·40	4·27	4·05	3·90	3·73	3·56
−1	3·38	3·30	3·17	3·06	2·97	2·88	2·79	2·74	2·66	2·51	4·49	4·51	4·50	4·46	4·39	4·27	4·05	3·91	3·74	3·57
−2	3·39	3·32	3·20	3·09	3·01	2·92	2·82	2·75	2·64	2·44	4·47	4·49	4·48	4·44	4·38	4·27	4·08	3·94	3·77	3·59
−3	3·41	3·34	3·22	3·12	3·04	2·95	2·84	2·75	2·62	2·41	4·45	4·46	4·45	4·41	4·36	4·27	4·11	3·97	3·81	3·61
−5	3·44	3·38	3·27	3·17	3·09	2·99	2·84	2·73	2·59	2·39	4·37	4·37	4·35	4·33	4·30	4·25	4·14	4·04	3·87	3·64
−10	3·49	3·43	3·33	3·22	3·13	3·01	2·83	2·71	2·56	2·37	4·20	4·19	4·18	4·17	4·16	4·14	4·11	4·06	3·96	3·71

$A = 1\cdot2$, $\mathbf{k} = 1\cdot0$

v	2·0	1·7	1·3	1·0	0·8	0·6	0·4	0·3	0·2	0·1	−2·0	−1·7	−1·3	−1·0	−0·8	−0·6	−0·4	−0·3	−0·2	−0·1
10	3·66	3·63	3·60	3·57	3·55	3·53	3·49	3·46	3·41	3·25	4·60	4·63	4·66	4·69	4·66	4·58	4·49	4·37	4·22	
5	3·52	3·46	3·39	3·33	3·29	3·24	3·18	3·13	3·04	2·84	4·71	4·73	4·76	4·76	4·74	4·68	4·55	4·43	4·26	4·07
3	3·45	3·37	3·24	3·14	3·07	3·00	2·92	2·86	2·77	2·58	4·76	4·79	4·82	4·81	4·78	4·70	4·52	4·37	4·18	3·96
2	3·43	3·33	3·17	3·03	2·94	2·84	2·74	2·68	2·60	2·43	4·79	4·82	4·85	4·84	4·80	4·70	4·50	4·33	4·13	3·90
1	3·41	3·31	3·12	2·96	2·83	2·70	2·56	2·49	2·42	2·30	4·81	4·84	4·88	4·87	4·82	4·71	4·48	4·30	4·09	3·85
0	3·41	3·30	3·11	2·93	2·79	2·64	2·48	2·40	2·31	2·22	4·81	4·85	4·89	4·88	4·83	4·72	4·48	4·29	4·07	3·84
−1	3·41	3·30	3·12	2·95	2·82	2·86	2·53	2·45	2·35	2·18	4·81	4·84	4·88	4·87	4·82	4·71	4·48	4·30	4·09	3·85
−2	3·42	3·32	3·15	2·99	2·87	2·83	2·58	2·49	2·35	2·13	4·78	4·81	4·83	4·82	4·77	4·68	4·49	4·33	4·12	3·88
−3	3·44	3·34	3·18	3·03	2·91	2·78	2·61	2·50	2·35	2·11	4·73	4·75	4·76	4·74	4·70	4·63	4·48	4·35	4·16	3·90
−5	3·47	3·38	3·23	3·09	2·97	2·74	2·64	2·50	2·33	2·09	4·61	4·61	4·59	4·56	4·54	4·50	4·43	4·34	4·19	3·91
−10	3·52	3·44	3·30	3·15	3·02	2·68	2·64	2·50	2·31	2·07	4·35	4·33	4·30	4·29	4·27	4·26	4·24	4·21	4·15	3·93

$A = 1\cdot2$, $\mathbf{k} = 0\cdot8$

v	2·0	1·7	1·3	1·0	0·8	0·6	0·4	0·3	0·2	0·1	−2·0	−1·7	−1·3	−1·0	−0·8	−0·6	−0·4	−0·3	−0·2	−0·1
10	3·67	3·65	3·63	3·61	3·60	3·58	3·55	3·53	3·48	3·32	4·42	4·46	4·47	4·48	4·48	4·45	4·36	4·27	4·15	4·00
5	3·54	3·50	3·44	3·40	3·37	3·33	3·28	3·24	3·16	2·96	4·50	4·52	4·54	4·54	4·51	4·46	4·32	4·21	4·06	3·87
3	3·48	3·41	3·31	3·23	3·18	3·12	3·06	3·01	2·93	2·74	4·54	4·57	4·59	4·57	4·54	4·46	4·29	4·15	3·98	3·78
2	3·45	3·37	3·24	3·14	3·06	2·99	2·91	2·86	2·79	2·63	4·56	4·59	4·61	4·60	4·56	4·46	4·27	4·12	3·93	3·73
1	3·44	3·35	3·20	3·07	2·97	2·86	2·75	2·70	2·64	2·53	4·58	4·61	4·63	4·62	4·57	4·46	4·25	4·09	3·90	3·68
0	3·43	3·34	3·19	3·04	2·93	2·82	2·69	2·62	2·55	2·49	4·58	4·61	4·64	4·63	4·58	4·47	4·25	4·08	3·88	3·67
−1	3·44	3·35	3·20	3·06	2·96	2·84	2·72	2·66	2·58	2·41	4·58	4·61	4·63	4·61	4·57	4·46	4·25	4·09	3·90	3·68
−2	3·45	3·36	3·22	3·10	3·00	2·89	2·77	2·69	2·57	2·36	4·56	4·58	4·59	4·58	4·53	4·45	4·27	4·13	3·93	3·71
−3	3·46	3·38	3·25	3·13	3·04	2·93	2·79	2·69	2·55	2·33	4·52	4·54	4·54	4·52	4·48	4·42	4·28	4·16	3·97	3·73
−5	3·49	3·42	3·30	3·19	3·09	2·97	2·81	2·69	2·53	2·31	4·43	4·43	4·41	4·39	4·37	4·33	4·26	4·18	4·03	3·77
−10	3·54	3·48	3·36	3·24	3·14	3·00	2·81	2·68	2·51	2·29	4·23	4·22	4·20	4·19	4·18	4·17	4·15	4·12	4·06	3·84

$A = 1\cdot2$, $\mathbf{k} = 0\cdot6$

v	2·0	1·7	1·3	1·0	0·8	0·6	0·4	0·3	0·2	0·1	−2·0	−1·7	−1·3	−1·0	−0·8	−0·6	−0·4	−0·3	−0·2	−0·1
10	3·70	3·69	3·67	3·66	3·65	3·64	3·62	3·60	3·55	3·40	4·25	4·26	4·28	4·29	4·27	4·24	4·14	4·05	3·93	3·78
5	3·58	3·55	3·51	3·48	3·46	3·44	3·40	3·36	3·29	3·09	4·31	4·32	4·33	4·32	4·29	4·23	4·10	3·99	3·85	3·68
3	3·53	3·47	3·40	3·34	3·30	3·26	3·21	3·17	3·10	2·91	4·34	4·35	4·36	4·34	4·31	4·23	4·07	3·94	3·78	3·60
2	3·50	3·44	3·35	3·26	3·20	3·15	3·09	3·05	2·98	2·82	4·35	4·37	4·38	4·36	4·31	4·22	4·05	3·91	3·74	3·55
1	3·49	3·42	3·31	3·21	3·13	3·05	2·96	2·92	2·86	2·76	4·36	4·38	4·39	4·37	4·32	4·22	4·03	3·88	3·71	3·52
0	3·49	3·42	3·30	3·19	3·10	3·01	2·91	2·86	2·81	2·75	4·36	4·38	4·40	4·37	4·32	4·22	4·03	3·88	3·71	3·50
−1	3·49	3·42	3·31	3·20	3·12	3·03	2·94	2·88	2·81	2·65	4·36	4·38	4·39	4·37	4·32	4·22	4·03	3·88	3·71	3·50
−2	3·50	3·43	3·33	3·23	3·15	3·07	2·97	2·90	2·79	2·59	4·34	4·36	4·37	4·34	4·30	4·22	4·05	3·92	3·75	3·54
−3	3·51	3·45	3·35	3·26	3·19	3·10	2·98	2·90	2·77	2·56	4·32	4·33	4·33	4·31	4·27	4·21	4·08	3·96	3·80	3·57
−5	3·54	3·48	3·39	3·30	3·23	3·13	2·99	2·89	2·74	2·53	4·26	4·26	4·24	4·23	4·21	4·17	4·10	4·03	3·88	3·63
−10	3·58	3·53	3·45	3·35	3·27	3·15	2·99	2·87	2·71	2·51	4·12	4·11	4·10	4·09	4·08	4·07	4·05	4·03	3·96	3·75

$A = 1\cdot0$, $\mathbf{k} = 1\cdot0$

v	2·0	1·7	1·3	1·0	0·8	0·6	0·4	0·3	0·2	0·1	−2·0	−1·7	−1·3	−1·0	−0·8	−0·6	−0·4	−0·3	−0·2	−0·1
10	3·72	3·70	3·68	3·66	3·65	3·63	3·61	3·59	3·56	3·45	4·40	4·43	4·46	4·49	4·51	4·50	4·44	4·37	4·25	4·09
5	3·61	3·57	3·52	3·47	3·45	3·41	3·37	3·33	3·27	3·10	4·49	4·51	4·54	4·56	4·56	4·53	4·43	4·33	4·18	3·98
3	3·56	3·50	3·40	3·31	3·26	3·21	3·14	3·10	3·03	2·85	4·53	4·56	4·60	4·61	4·60	4·55	4·42	4·29	4·11	3·88
2	3·54	3·46	3·34	3·22	3·14	3·06	2·98	2·93	2·86	2·69	4·55	4·59	4·63	4·64	4·62	4·56	4·41	4·26	4·06	3·82
1	3·53	3·44	3·29	3·15	3·04	2·92	2·79	2·73	2·66	2·54	4·57	4·61	4·67	4·67	4·65	4·58	4·40	4·23	4·02	3·76
0	3·52	3·44	3·28	3·12	3·00	2·86	2·71	2·63	2·54	2·46	4·58	4·62	4·67	4·68	4·66	4·58	4·39	4·22	4·00	3·74
−1	3·53	3·44	3·29	3·14	3·03	2·90	2·75	2·67	2·58	2·41	4·57	4·61	4·65	4·66	4·15	4·57	4·39	4·23	4·02	3·76
−2	3·54	3·46	3·32	3·18	3·07	2·95	2·81	2·72	2·59	2·36	4·55	4·58	4·61	4·61	4·36	4·53	4·39	4·26	4·06	3·80
−3	3·55	3·47	3·34	3·22	3·12	2·99	2·84	2·73	2·58	2·34	4·51	4·53	4·54	4·53	4·51	4·46	4·37	4·28	4·11	3·83
−5	3·57	3·51	3·39	3·27	3·17	3·04	2·86	2·74	2·56	2·31	4·41	4·41	4·40	4·38	4·59	4·33	4·29	4·24	4·14	3·87
−10	3·61	3·55	3·44	3·32	3·21	3·07	2·87	2·73	2·55	2·30	4·21	4·19	4·18	4·16	4·64	4·14	4·13	4·12	4·08	3·94

$A = 1\cdot0$, $\mathbf{k} = 0\cdot8$

v	2·0	1·7	1·3	1·0	0·8	0·6	0·4	0·3	0·2	0·1	−2·0	−1·7	−1·3	−1·0	−0·8	−0·6	−0·4	−0·3	−0·2	−0·1
10	3·73	3·72	3·70	3·69	3·68	3·67	3·65	3·64	3·61	3·50	4·27	4·29	4·32	4·34	4·34	4·33	4·26	4·19	4·07	3·90
5	3·63	3·60	3·56	3·53	3·51	3·48	3·45	3·42	3·36	3·19	4·33	4·35	4·37	4·38	4·38	4·34	4·25	4·15	4·00	3·81
3	3·58	3·53	3·45	3·39	3·34	3·30	3·25	3·21	3·15	2·97	4·37	4·39	4·42	4·42	4·41	4·36	4·23	4·11	3·94	3·72
2	3·56	3·50	3·40	3·30	3·24	3·17	3·11	3·07	3·00	2·85	4·38	4·41	4·44	4·44	4·42	4·36	4·21	4·08	3·89	3·67
1	3·55	3·48	3·36	3·24	3·15	3·05	2·95	2·90	2·84	2·73	4·40	4·43	4·46	4·47	4·44	4·37	4·20	4·05	3·85	3·62
0	3·54	3·47	3·35	3·22	3·12	3·00	2·88	2·81	2·74	2·67	4·40	4·43	4·47	4·47	4·45	4·37	4·19	4·04	3·83	3·60
−1	3·55	3·48	3·36	3·23	3·14	3·03	2·92	2·85	2·77	2·60	4·40	4·42	4·46	4·46	4·44	4·37	4·07	4·05	3·85	3·62
−2	3·56	3·49	3·38	3·27	3·18	3·08	2·96	2·88	2·76	2·55	4·38	4·40	4·42	4·42	4·40	4·34	4·17	4·09	3·90	3·66
−3	3·57	3·51	3·40	3·30	3·22	3·12	2·98	2·89	2·75	2·52	4·35	4·36	4·36	4·34	4·30	4·21	4·12	3·96	3·69	
−5	3·59	3·54	3·44	3·35	3·26	3·15	3·00	2·89	2·73	2·50	4·28	4·28	4·27	4·25	4·23	4·21	4·21	4·13	4·02	3·76
−10	3·63	3·58	3·49	3·39	3·30	3·18	3·01	2·88	2·71	2·48	4·12	4·11	4·10	4·09	4·09	4·08	4·20	4·05	4·02	3·87

Table 5.5 (continued)

$A = 1 \cdot 0, \; k = 0 \cdot 6$

	2·0	1·7	1·3	1·0	0·8	0·6	0·4	0·3	0·2	0·1	−2·0	−1·7	−1·3	−1·0	−0·8	−0·6	−0·4	−0·3	−0·2	−0·1
10	3·75	3·74	3·73	3·72	3·72	3·71	3·70	3·68	3·66	3·56	4·15	4·16	4·18	4·19	4·18	4·16	4·08	4·01	3·89	3·72
5	3·66	3·64	3·61	3·59	3·57	3·55	3·53	3·50	3·45	3·29	4·19	4·20	4·22	4·22	4·20	4·16	4·06	3·97	3·83	3·64
3	3·62	3·58	3·52	3·47	3·44	3·41	3·37	3·33	3·28	3·11	4·21	4·23	4·24	4·24	4·22	4·17	4·04	3·93	3·77	3·56
2	3·60	3·55	3·47	3·40	3·35	3·30	3·25	3·21	3·15	3·00	4·22	4·24	4·26	4·25	4·23	4·17	4·02	3·90	3·72	3·52
1	3·59	3·54	3·44	3·35	3·28	3·20	3·12	3·07	3·02	2·92	4·23	4·25	4·27	4·27	4·24	4·17	4·00	3·87	3·69	3·48
0	3·59	3·53	3·43	3·33	3·25	3·16	3·06	3·01	2·95	2·90	4·24	4·26	4·28	4·27	4·24	4·17	4·01	3·87	3·69	3·48
−1	3·59	3·54	3·44	3·35	3·27	3·18	3·09	3·03	2·96	2·80	4·23	4·25	4·27	4·27	4·24	4·17	4·01	3·87	3·69	3·48
−2	3·60	3·55	3·46	3·37	3·30	3·22	3·12	3·05	2·95	2·74	4·22	4·24	4·25	4·24	4·21	4·16	4·03	3·92	3·74	3·52
−3	3·61	3·56	3·48	3·40	3·33	3·25	3·14	3·06	2·93	2·71	4·20	4·21	4·22	4·20	4·18	4·14	4·05	3·96	3·81	3·56
−5	3·63	3·59	3·51	3·44	3·37	3·28	3·15	3·05	2·90	2·68	4·15	4·15	4·14	4·13	4·12	4·10	4·06	4·01	3·90	3·65
−10	3·66	3·63	3·56	3·48	3·40	3·30	3·15	3·04	2·88	2·66	4·05	4·04	4·03	4·03	4·02	4·01	4·00	3·99	3·96	3·81

On the other hand, when kT/λ is negative, reducing **k** merely reduces $\bar{\mu}$ at all values of v, A and kT/λ.

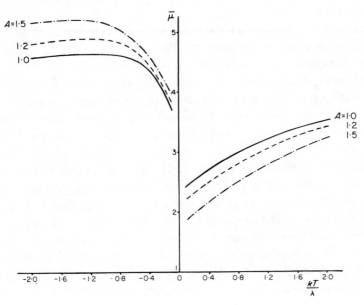

Figure 5.13. The variation of $\bar{\mu}$ *vs.* kT/λ for the $^4 T_1$ term for various values of A, with $v = 0$ and **k** = 1·0.

(c) *The variation of $\bar{\mu}$ with A* (illustrated in Fig. 5.13). As in the previous case the general trends depend upon the sign of kT/λ. When kT/λ is positive, varying A from 1·5 to 1·0 increases $\bar{\mu}$ at all values of v, **k**, and kT/λ. With kT/λ negative the usual tendency is to reduce $\bar{\mu}$ as A is reduced at all values of v, **k** and kT/λ except with very high values of v combined with low values of kT/λ.

5.3. The magnetic properties of A_2 and E terms in axially symmetric crystal fields

One of the most important effects of lowering the symmetry from cubic is to remove some of the orbital degeneracy of orbital triplet ground terms. This loss of orbital degeneracy leads to a reduction in the orbital contribution to the magnetic susceptibility which therefore tends towards the spin-only value. However, there is no orbital contribution to the susceptibility directly from the A_2 and E terms discussed in Chapter 4 and thus even in the case of the E terms, where lowering of the symmetry may remove the orbital degeneracy, the susceptibility to a first approximation should still be the spin-only value. Anticipating some of the results which will be calculated later in this section, we find that just as in the cubic field case the excited terms may be mixed with the ground term. This mixing not only makes the susceptibility different from the spin-only value but also makes it anisotropic.

The mixing between the ground and excited terms may be caused by the following mechanisms:

(a) If any of the excited terms transform in the same way as the ground term in the point group of the molecule, then the crystal field will cause mixing between the ground and excited term wave-functions.

(b) Spin–orbit coupling may cause mixing between excited and ground term functions, just as in the cubic field case. Such mixing is inversely proportional to the energy separation between the terms being mixed and thus splittings in the excited terms now cause differing degrees of admixture compared with the cubic field case.

The magnetic anisotropy is now a reflection of the magnitude and often also the sense of the splittings in the excited terms (i.e. their relative ordering). Another effect of the combined action of the crystal field and spin–orbit coupling occurs when there is more than one unpaired electron in the system and that is that some of the spin degeneracy of the ground term may be removed even before the application of a magnetic field. This phenomenon, known as 'zero-field splitting', can, if it becomes large, have an important effect on the temperature variation of the magnetic susceptibility and also on the electron spin resonance properties of the system.

The relationship between magnetic anisotropy, zero-field splittings, and excited term splittings will now be illustrated by calculations on a few simple examples.

5.3.1. THE ^2E_g TERM FROM THE d^9 CONFIGURATION IN A TETRAGONALLY DISTORTED OCTAHEDRAL ENVIRONMENT

Two different situations will be considered here. The first corresponds to that of an elongation along the four-fold axis which leaves the $d_{x^2-y^2}$ *orbital* at the highest energy and it therefore contains the unpaired electron. The second situation is that of a tetragonal compression which would leave the unpaired electron in the d_{z^2} *orbital*.

Tetragonal elongation

The energy diagram for this situation is shown in Fig. 5.14. For the purpose of illustrating the effects of the distortion on the magnetic properties of this system we will confine our attention to the situation where the crystal field splittings are very much greater than spin–orbit coupling or the thermal energy available to the system.

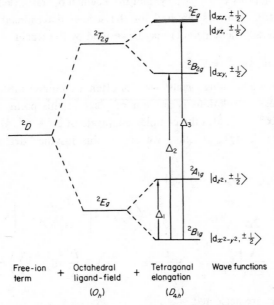

Free-ion term + Octahedral ligand-field (O_h) + Tetragonal elongation (D_{4h}) — Wave functions

Figure 5.14. The splitting diagram of the ^2D term in a tetragonally elongated octahedral environment.

Although spin–orbit coupling does not remove any of the degeneracy of the $^2B_{1g}$ ground state it mixes excited state wave-functions into the ground state. The corrected wave-functions for each of the ground state components obtained by using Equation 3.25 and Table 4.1 are:

$$|x^2 - y^2, \tfrac{1}{2}\rangle = |d_{x^2-y^2}, \tfrac{1}{2}\rangle - \frac{\lambda}{2\Delta_3}|d_{xz}, -\tfrac{1}{2}\rangle + \frac{i\lambda}{2\Delta_3}|d_{yz}, -\tfrac{1}{2}\rangle$$
$$+ \frac{i\lambda}{\Delta_2}|d_{xy}, \tfrac{1}{2}\rangle$$

$$|x^2 - y^2, -\tfrac{1}{2}\rangle = |d_{x^2-y^2}, -\tfrac{1}{2}\rangle + \frac{\lambda}{2\Delta_3}|d_{xz}, \tfrac{1}{2}\rangle + \frac{i\lambda}{2\Delta_3}|d_{yz}, \tfrac{1}{2}\rangle$$
$$+ \frac{i\lambda}{\Delta_2}|d_{xy}, -\tfrac{1}{2}\rangle$$

$$. (5.25)$$

These functions may be regarded as normalized if $\Delta_i \gg \lambda$ since terms in $(\lambda/\Delta_i)^2$ will be negligible compared with unity. The above corrected ground state wave-functions are now used to calculate first- and second-order Zeeman coefficients.

(a) The first- and second-order Zeeman coefficients for the molecular z-direction. The first-order Zeeman coefficients are obtained from the secular determinant involving the functions $|x^2 - y^2, \pm\frac{1}{2}\rangle$ and the Zeeman operator $(k_z\hat{l}_z + 2\hat{s}_z)\beta H_z$. One of the matrix elements required for the secular determinant will now be worked out in detail, the remainder may be verified by the reader.

Example $\langle x^2 - y^2, \frac{1}{2}|k_z\hat{l}_z + 2\hat{s}_z|x^2 - y^2, \frac{1}{2}\rangle$

Before proceeding a warning against what is often a common mistake made by beginners in these calculations is not out of place at this point. It is vital to remember that $\langle x^2 - y^2, \frac{1}{2}|$ is the complex conjugate of $|x^2 - y^2, \frac{1}{2}\rangle$ and that for example $d^*_{x^2-y^2}$ is $d_{x^2-y^2}$ since we are in this instance dealing with *real* orbitals. Thus:

$$\langle x^2 - y^2, \frac{1}{2}|k_z\hat{l}_z + 2\hat{s}_z|x^2 - y^2, \frac{1}{2}\rangle$$

$$= \left(\left(\langle d_{x^2-y^2}, \frac{1}{2}| - \frac{\lambda}{2\Delta_3} \langle d_{xz}, -\frac{1}{2}| - \frac{i\lambda}{2\Delta_3} \langle d_{yz}, -\frac{1}{2}| + \frac{i\lambda}{\Delta_2} \langle d_{xy}, \frac{1}{2}| \right) k_z\hat{l}_z + 2\hat{s}_z \right|$$

$$\left| d_{x^2-y^2}, \frac{1}{2} \rangle - \frac{\lambda}{2\Delta_3}|d_{xz}, -\frac{1}{2}\rangle + \frac{i\lambda}{2\Delta_3}|d_{yz}, -\frac{1}{2}\rangle - \frac{i\lambda}{\Delta_2}|d_{xy}, \frac{1}{2}\rangle \right) = 1 - \frac{4k_z\lambda}{\Delta_2}$$

if terms in $(\lambda/\Delta)^2$ are neglected.

The required secular determinant is:

	$|x^2 - y^2, \frac{1}{2}\rangle$	$|x^2 - y^2, -\frac{1}{2}\rangle$
$\langle x^2 - y^2, \frac{1}{2}|$	$\left(1 - \dfrac{4k_z\lambda}{\Delta_2}\right)\beta H_z - E$	0
$\langle x^2 - y^2, -\frac{1}{2}|$	0	$-\left(1 - \dfrac{4k_z\lambda}{\Delta_2}\right)\beta H_z - E$

$= 0.$

$$(5.26)$$

which leads to $E = \pm (1 - 4k_z\lambda/\Delta_2)\beta H_z$. This means that the first-order Zeeman coefficients are $W^{(1)}(z) = \pm (1 - 4k_z\lambda/\Delta_2)\beta$ and the spectroscopic splitting factor $g_z = 2(1 - 4k_z\lambda/\Delta_2)$.

The second-order Zeeman coefficient for each of the ground state components is calculated using Equation 3.26. This equation involves the excited state wave-functions and their energies and in a more exact calculation it would be necessary to calculate the effect of spin–orbit coupling on the excited states as well as the ground term. However, we have assumed that $\lambda \ll \Delta$ in this particular example and thus it is within this approximation to use the excited state wave-functions and energies given in Fig. 5.14. Thus for each of the ground state components the second order Zeeman coefficient is

$$W^{(2)}(z) = \sum_i \frac{\beta^2 \langle x^2 - y^2, \tfrac{1}{2} | k_z \hat{l}_z + 2\hat{s}_z | \psi_i \rangle \langle \psi_i | k_z \hat{l}_z + 2\hat{s}_z | x^2 - y^2, \tfrac{1}{2} \rangle}{E(x^2 - y^2, \tfrac{1}{2}) - E_i^0}$$

$$= -\frac{4k_z^2 \beta^2}{\Delta_2},$$

where i refers to the excited states and again terms in $(\lambda/\Delta)^2$ are ignored.

With the assumption that $\Delta \gg kT$ the susceptibility in the z or parallel direction becomes

$$\chi_\| = \frac{N\beta^2}{3kT} 3 \left(1 - \frac{4k_z\lambda}{\Delta_2} \right)^2 + \frac{8N\beta^2 k_z^2}{\Delta_2}. \tag{5.27}$$

(b) *The magnetic susceptibility and g value for the perpendicular direction.* Since the system has axial symmetry we need only calculate the result for the molecular x-direction, that for the y-direction being identical. The calculation is the same as for the parallel direction with $(k_x \hat{l}_x + 2\hat{s}_x)\beta H_x$ as the Zeeman operator. The results now are

$$W^{(1)}(x) = \pm \left(1 - \frac{\lambda k_x}{\Delta_3} \right), \quad g_x = 2 \left(1 - \frac{\lambda k_x}{\Delta_3} \right),$$

and

$$W^{(2)}(x) = -\frac{k_x^2 \beta^2}{\Delta_3},$$

which leads to the magnetic susceptibility

$$\chi_\perp = \frac{N\beta^2}{3kT} 3 \left(1 - \frac{k_x\lambda}{\Delta_3} \right)^2 + \frac{2N\beta^2 k_x^2}{\Delta_3}. \tag{5.28}$$

The magnetic susceptibilities and the g values for the system are anisotropic. It should be noted that, although the values of $\chi_\|$ and $\chi_\perp$ will be different from each other, they should still obey the Curie-law plus a T.I.P. contribution as in the cubic field situation. From the form of the susceptibility or g value equations we can see that a determination of these quantities gives some information concerning the energies of the excited states. However, more importantly, if we can measure the energies of the excited states from the electronic absorption spectrum of the compound, the principal magnetic susceptibilities or g values enable k or, if we had chosen to express our wave-functions as molecular orbitals, molecular orbital coefficients to be estimated.

Tetragonal compression

The most likely energy diagram for this situation is given in Fig. 5.15.

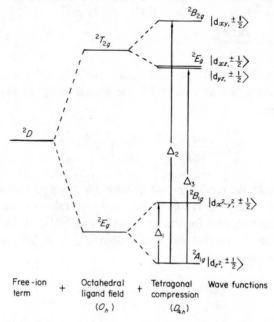

Figure 5.15. The splitting diagram of the ^2D term in a tetragonally compressed octahedral environment.

Proceeding in the same manner and making the same assumptions as in the case of the $^2B_{1g}$ ground state we find:

$$|z^2, \pm \tfrac{1}{2}\rangle = |d_{z^2} \pm \tfrac{1}{2}\rangle + \frac{i\sqrt{3}\lambda}{2\Delta_3}|d_{yz}, \mp \tfrac{1}{2}\rangle \pm \frac{\sqrt{3}\lambda}{2\Delta_3}|d_{xz}, \mp \tfrac{1}{2}\rangle, \qquad (5.29)$$

$$g_z = 2 \quad \text{and} \quad g_x = 2\left(1 - \frac{3k_x\lambda}{\Delta_3}\right), \qquad (5.30)$$

$$\chi_\parallel = \frac{N\beta^2}{3kT} \cdot 3, \qquad (5.31)$$

$$\chi_\perp = \frac{N\beta^2}{3kT} 3\left(1 - \frac{3k_x\lambda}{\Delta_3}\right)^2 + \frac{6N\beta^2 k_x^2}{\Delta_3}. \qquad (5.32)$$

As expected, the magnetic properties are anisotropic, but a very important feature is that for this $^2A_{1g}$ ground state the magnetic properties in the parallel direction correspond to spin-only values. This observation should provide a useful means of deciding when the $^2A_{1g}$ term is the ground state in these systems. The reason for the spin-only properties in the parallel molecular direction is that the z-component of the orbital angular momentum operator does not connect any of the other functions with d_{z^2} and hence there is no mechanism whereby an orbital contribution may be introduced into the ground state when the parallel direction is being considered.

5.3.2. THE $^4A_{2g}$ TERM FROM THE d^7 CONFIGURATION IN A TETRAGONALLY DISTORTED TETRAHEDRAL ENVIRONMENT

This particular example has been chosen to illustrate the possible occurence of zero-field splittings in orbitally singly degenerate ground terms which arise from multi-electron configurations. Although we will make some simplifying assumptions it provides a good approximation to the magnetic behaviour of many 'tetrahedral' cobalt(II) complexes which often show small angular distortions.

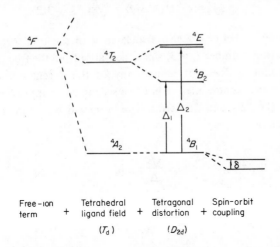

Figure 5.16. The partial splitting diagram of the ^4F term under the action of a tetragonally distorted tetrahedral crystal field and spin–orbit coupling.

For our present purposes we will make the assumptions that we need only consider the interactions between the components of the cubic field $^4A_2$ and $^4T_2$ terms and that $\Delta \gg \lambda$ and kT. The wave-functions, in terms of $|M_L, M_S\rangle$, appropriate to this calculation are:

$$^4B_1 \quad \phi_1 = \sqrt{\tfrac{1}{2}}(|2, 3/2\rangle - |-2, 3/2\rangle)$$

$$\phi_2 = \sqrt{\tfrac{1}{2}}(|2, \tfrac{1}{2}\rangle - |-2, \tfrac{1}{2}\rangle)$$

$$\phi_3 = \sqrt{\tfrac{1}{2}}(|2, -\tfrac{1}{2}\rangle - |-2, -\tfrac{1}{2}\rangle)$$

$$\phi_4 = \sqrt{\tfrac{1}{2}}(|2, -3/2\rangle - |-2, -3/2\rangle)$$

$$^4B_2 \quad \phi_5 = \sqrt{\tfrac{1}{2}}(|2, 3/2\rangle + |-2, 3/2\rangle)$$

$$\phi_6 = \sqrt{\tfrac{1}{2}}(|2, \tfrac{1}{2}\rangle + |-2, \tfrac{1}{2}\rangle)$$

$$\phi_7 = \sqrt{\tfrac{1}{2}}(|2, -\tfrac{1}{2}\rangle + |-2, -\tfrac{1}{2}\rangle)$$

$$\phi_8 = \sqrt{\tfrac{1}{2}}(|2, -3/2\rangle + |-2, -3/2\rangle)$$

$$^4E \quad \phi_9 = \sqrt{\tfrac{5}{8}}|-1, 3/2\rangle - \sqrt{\tfrac{3}{8}}|3, 3/2\rangle$$

$$\phi_{10} = \sqrt{\tfrac{5}{8}}|-1, \tfrac{1}{2}\rangle - \sqrt{\tfrac{3}{8}}|3, \tfrac{1}{2}\rangle$$

$$\phi_{11} = \sqrt{\tfrac{5}{8}}|-1, -\tfrac{1}{2}\rangle - \sqrt{\tfrac{3}{8}}|3, -\tfrac{1}{2}\rangle$$

$$\phi_{12} = \sqrt{\tfrac{5}{8}}|-1, -3/2\rangle - \sqrt{\tfrac{3}{8}}|3, -3/2\rangle$$

$$\phi_{13} = \sqrt{\tfrac{5}{8}}|1, 3/2\rangle - \sqrt{\tfrac{3}{8}}|-3, 3/2\rangle$$

$$\phi_{14} = \sqrt{\tfrac{5}{8}}|1, \tfrac{1}{2}\rangle - \sqrt{\tfrac{3}{8}}|-3, \tfrac{1}{2}\rangle$$

$$\phi_{15} = \sqrt{\tfrac{5}{8}}|1, -\tfrac{1}{2}\rangle - \sqrt{\tfrac{3}{8}}|-3, -\tfrac{1}{2}\rangle$$

$$\phi_{16} = \sqrt{\tfrac{5}{8}}|1, -3/2\rangle - \sqrt{\tfrac{3}{8}}|-3, -3/2\rangle.$$

Each of the above functions is also associated with the total orbital angular momentum quantum number $L = 3$, since they arise from the free ion ^4F term.

Spin–orbit coupling does not remove any of the degeneracy of the $^4B_1$ ground term to a first approximation but to a second-order approximation there is mixing with the $^4B_2$ and ^4E terms. The corrected wave-functions for the ground term are now:

$$\Psi_1 = \phi_1 - \frac{3\lambda}{\Delta_1}\phi_5 + \frac{\sqrt{6}\lambda}{\Delta_2}\phi_{10}$$

$$\Psi_2 = \phi_2 - \frac{\lambda}{\Delta_1}\phi_6 + \frac{2\sqrt{2}\lambda}{\Delta_2}\phi_{11} - \frac{\sqrt{6}\lambda}{\Delta_2}\phi_{13}$$

$$\Psi_3 = \phi_3 + \frac{\lambda}{\Delta_1}\phi_7 - \frac{2\sqrt{2}\lambda}{\Delta_2}\phi_{14} + \frac{\sqrt{6}\lambda}{\Delta_2}\phi_{12}$$

$$\Psi_4 = \phi_4 + \frac{3\lambda}{\Delta_1}\phi_8 - \frac{\sqrt{6}\lambda}{\Delta_2}\phi_{15}$$

(5.33)

assuming $(\lambda/\Delta)^2 \ll 1$. Δ_1 and Δ_2 are defined in Fig. 5.16.

The energies of the ground state components after spin–orbit coupling, calculated using Equation 3.26 are:

$$E_1 = E_4 = E_1^0 - \frac{9\lambda^2}{\Delta_1} - \frac{6\lambda^2}{\Delta_2}$$

$$E_2 = E_3 = E_2^0 - \frac{\lambda^2}{\Delta_1} - \frac{14\lambda^2}{\Delta_2} \qquad (5.34)$$

We now find that, unless $\Delta_1 = \Delta_2$, i.e. the undistorted situation, some of the degeneracy of the ground term is removed. This zero-field splitting corresponds to the removal of some of the spin-degeneracy of the $^4B_1$ term since the new functions are primarily specified by $M_S = \pm\frac{3}{2}$ at E_1 and $M_S = \pm\frac{1}{2}$ at E_2. The value of δ, defined as $\delta = E_2 - E_1$, is

$$8\lambda^2 \left(\frac{1}{\Delta_1} - \frac{1}{\Delta_2} \right)$$

and obviously its magnitude and sign are very dependent on the excited state splittings.

The magnetic susceptibilities may now be calculated from the wave-functions (5.33). The matrices of these functions for the magnetic moment operator in the x and z direction are:

$\hat{\mu}_z$

	Ψ_1	Ψ_2	Ψ_3	Ψ_4
Ψ_1	$3\left(1 - \dfrac{4k\lambda}{\Delta_1}\right)\beta H_z$	0	0	0
Ψ_2	0	$\left(1 - \dfrac{4k\lambda}{\Delta_1}\right)\beta H_z$	0	0
Ψ_3	0	0	$-\left(1 - \dfrac{4k\lambda}{\Delta_1}\right)\beta H_z$	0
Ψ_4	0	0	0	$-3\left(1 - \dfrac{4k\lambda}{\Delta_1}\right)\beta H_z$

$$(5.35)$$

$\hat{\mu}_x$

	Ψ_1	Ψ_2	Ψ_3	Ψ_4
Ψ_1	0	$\sqrt{3}\left(1 - \dfrac{4k\lambda}{\Delta_2}\right)\beta H_x$	0	0
Ψ_2	$\sqrt{3}\left(1 - \dfrac{4k\lambda}{\Delta_2}\right)\beta H_x$	0	$2\left(1 - \dfrac{4k\lambda}{\Delta_2}\right)\beta H_x$	0
Ψ_3	0	$2\left(1 - \dfrac{4k\lambda}{\Delta_2}\right)\beta H_x$	0	$\sqrt{3}\left(1 - \dfrac{4k\lambda}{\Delta_2}\right)\beta H_x$
Ψ_4	0	0	$\sqrt{3}\left(1 - \dfrac{4k\lambda}{\Delta_2}\right)\beta H_x$	0

$$(5.36)$$

Since the matrix for $\hat{\mu}_z$ is diagonal the calculation of χ_z presents no problem, the same result being obtained whether δ is comparable to or greater than the energy changes in the magnetic field. Including the second-order Zeeman contribution from the $^4B_2$ term, with the assumption that δ may be ignored compared with Δ_1, the expression is

$$\chi_z = \frac{N\beta^2}{3kT} 3 \left[\frac{18\left(1 - \frac{4k\lambda}{\Delta_1}\right)^2 + 2\left(1 - \frac{4k\lambda}{\Delta_1}\right)^2 \exp\left(-\frac{\delta}{kT}\right)}{2 + 2\exp\left(-\delta/kT\right)} \right] + \frac{8N\beta^2 k^2}{\Delta_1}$$

For $\delta \ll kT$, $\exp\left(-\delta/kT\right) \approx 1 - \delta/kT + \frac{1}{2}(\delta/kT)^2$, this equation becomes:

$$\chi_z = \frac{N\beta^2}{3kT}\left[15\left(1 - \frac{4k\lambda}{\Delta_1}\right)^2 \left(1 + \frac{2\delta}{5kT}\right)\right] + \frac{8N\beta^2 k^2}{\Delta_1}. \tag{5.37}$$

Because of the off-diagonal elements in the $\hat{\mu}_x$ matrix which connect functions which are not degenerate with each other in the absence of the magnetic field, the calculation of χ_x may be more complicated. If we consider the effects of spin–orbit coupling and the magnetic field as simultaneous perturbations on the $^4B_1$ ground term then the secular determinant to be solved is (where the energy of Ψ_1 and Ψ_4 in the absence of a magnetic field is defined as zero)

	$|\Psi_1\rangle$	$|\Psi_2\rangle$	$|\Psi_3\rangle$	$|\Psi_4\rangle$	
$\langle\Psi_1|$	$0 - E$	$\sqrt{3}xH_x$	0	0	
$\langle\Psi_2|$	$\sqrt{3}xH_x$	$\delta - E$	$2xH_x$	0	$= 0.$
$\langle\Psi_3|$	0	$2xH_x$	$\delta - E$	$\sqrt{3}xH_x$	
$\langle\Psi_4|$	0	0	$\sqrt{3}xH_x$	$0 - E$	

$$\tag{5.38}$$

where

$$x = \left(1 - \frac{4k\lambda}{\Delta_2}\right)\beta.$$

Equation 5.38 is a fourth-order equation in E and as such is usually difficult to solve. At this stage there are two alternatives we can take:

(a) the equation could be solved either algebraically or numerically, whichever is the most convenient, and expressions found for the variation of the energy levels with magnetic field;

(b) make the assumption that the zero-field splitting is greater than the effect of the magnetic field and treat the latter as a perturbation.

Although (a) is the more difficult approach, we will use it because of its generality.

Expansion of (5.38) gives:

$$E^4 - 2\delta E^3 - (10x^2 H^2 - \delta^2)E^2 + 6\delta x^2 H^2 E + 9x^4 H^4 = 0, \qquad (5.39)$$

where

$$x = \left(1 - \frac{4k\lambda}{\Delta_2}\right)\beta.$$

Fortunately Equation 5.39 factorizes giving:

$$(E^2 - E(2xH + \delta) - 3x^2 H^2)(E^2 + E(2xH - \delta) - 3x^2 H^2) = 0,$$

leading to the energy expressions:

$$\left.\begin{aligned}
E_1 &= \tfrac{1}{2}[2xH + \delta - (16x^2 H^2 + 4xH\delta + \delta^2)^{\frac{1}{2}}] \\
E_2 &= \tfrac{1}{2}[2xH + \delta + (16x^2 H^2 + 4xH\delta + \delta^2)^{\frac{1}{2}}] \\
E_3 &= \tfrac{1}{2}[-2xH + \delta + (16x^2 H^2 - 4xH\delta + \delta^2)^{\frac{1}{2}}] \\
E_4 &= \tfrac{1}{2}[-2xH + \delta - (16x^2 H^2 - 4xH\delta + \delta^2)^{\frac{1}{2}}]
\end{aligned}\right\}. \qquad (5.40)$$

These equations are not in the form of the power series required for substitution into Van Vleck's susceptibility equation, but can be transformed by the application of Maclaurin's theorem giving, to terms in H^2:

$$\left.\begin{aligned}
E_1 &= -\frac{3x^2}{\delta} \cdot H^2 \\[2mm]
E_2 &= \delta + 2xH + \frac{3x^2}{\delta} \cdot H^2 \\[2mm]
E_3 &= \delta - 2xH + \frac{3x^2}{\delta} \cdot H^2 \\[2mm]
E_4 &= -\frac{3x^2}{\delta} \cdot H^2
\end{aligned}\right\}. \qquad (5.41)$$

The first- and second-order Zeeman coefficients for levels E_1 to E_4 are now readily obtained and substitution into Van Vleck's equation gives:

$$\chi_x = \frac{N\beta^2}{3kT} 3\left[\frac{(4x^2 - (6kTx^2/\delta))\exp(-\delta/kT) + (6kTx^2/\delta)}{\exp(-\delta/kT) + 1}\right] + \frac{8N\beta^2 k^2}{\Delta_2}$$

where

$$x = \left(1 - \frac{4k\lambda}{\Delta_2}\right)$$

With the approximation that

$$\exp(-\delta/kT) = 1 - \frac{\delta}{kT} + \frac{1}{2}\left(\frac{\delta}{kT}\right)^2$$

this equation becomes:

$$\chi_x = \frac{N\beta^2}{3kT}\left[15x^2\left(1 - \frac{\delta}{5kT}\right)\right] + \frac{8N\beta^2 k^2}{\Delta_2}. \qquad (5.42)$$

Combining (5.42) and (5.37) gives the following expression for the average susceptibility:

$$\bar{\chi} = \frac{N\beta^2}{3kT}\left[5y^2 + 10x^2 + \frac{2\delta}{kT}(y^2 - x^2)\right] + \frac{8N\beta^2 k^2}{3}\left[\frac{1}{\Delta_1} + \frac{2}{\Delta_2}\right], \qquad (5.43)$$

where

$$y = \left(1 - \frac{4k\lambda}{\Delta_1}\right)$$

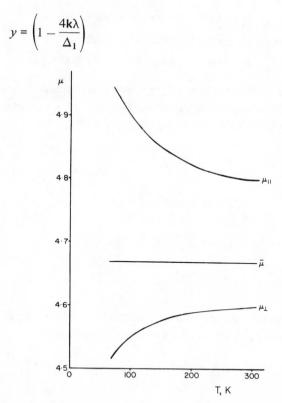

Figure 5.17. The variation of μ vs. T for the $^4A_2$ term (from the d^7 configuration) in a tetragonally distorted tetrahedral crystal field when the T.I.P. term is neglected and $\lambda = -170$ cm^{-1}, $k = 1 \cdot 0$, $\Delta_1 = 3000$ cm^{-1}, $\Delta_2 = 3500$ cm^{-1}, and $\delta = 11$ cm^{-1}.

The effect of the zero-field splitting on the magnetic behaviour is conveniently illustrated in terms of μ. Figure 5.17 shows the variation of μ with temperature predicted from Equations 5.37, 5.42, and 5.43, *when the T.I.P. term is neglected*, for the parameters $\lambda = -170$ cm^{-1}, $\mathbf{k} = 1{\cdot}0$, $\Delta_1 = 3000$ cm^{-1}, $\Delta_2 = 3500$ cm^{-1}, and $\delta = 11$ cm^{-1}. Although $\bar{\mu}$ is constant over the temperature range 75–300 K the values of $\mu_\|$ and $\mu_\perp$ diverge considerably. The temperature variation of $\mu_\|$ and $\mu_\perp$ is a consequence of the term in δ in Equations 5.37, 5.42, and 5.43, which reflects the changing thermal population of the ground state components. The reason for the fall in $\mu_\perp$ with temperature in this particular example may be correlated with the fact that there is thermal depopulation into components which have no first-order Zeeman effect associated with them in the x-direction. Thus in the limit $T \to 0$, the susceptibility will be independent of temperature and hence $\mu_\perp \propto \sqrt{T}$. If δ has a negative sign the temperature variation of $\mu_\|$ and $\mu_\perp$ would be in the reverse sense to that shown in Fig. 5.17 and in principle the measurement of the anisotropy should determine the sign of δ.

Concluding remarks

In this chapter we have shown some of the effects on the magnetic properties of axially distorting octahedral and tetrahedral complexes. In the case where there were orbital triplet ground terms in cubic symmetry, we have restricted the models to a consideration of the effects of the lower symmetry and spin–orbit coupling on these ground terms only. This assumes that the contributions to the magnetic properties from the excited terms are negligible compared with those from the orbital triplet ground term. The application and a discussion of the usefulness of these models in some real systems is given in Chapter 6.

REFERENCES

1. J.E. Griffiths, J. Owen and I.M. Ward (1953). *Proc. Roy. Soc. (London)*, **A226**, 96; K.W.H. Stevens (1953). *Proc. Roy. Soc. (London)*, **A219**, 542.
2. J. Owen and J.H.M. Thornley (1966). *Reports Progr. Phys.*, **29**, 675; M. Gerloch and J.R. Miller (1968). *Progr. Inorg. Chem.*, **10**, 1.
3. B.N. Figgis (1961). *Trans. Farad. Soc.*, **57**, 198.
4. B.N. Figgis, J. Lewis, F.E. Mabbs and G.A. Webb (1966). *J. Chem. Soc. (A)*, 1411.
5. B.N. Figgis, J. Lewis, F.E. Mabbs and G.A. Webb (1967). *J. Chem. Soc. (A)*, 442.
6. B.N. Figgis, M. Gerloch, J. Lewis, F.E. Mabbs and G.A. Webb (1968). *J. Chem. Soc. (A)*, 2086.

A comparison of experimental and calculated data

In this chapter the various models for the behaviour of the magnetic susceptibilities already outlined in Chapters 4 and 5 will be applied to a number of selected compounds. It is not our intention to give a comprehensive survey of known magnetic results, but merely to examine how well the previous models account for the observed data on a few selected examples. Where the models break down a summary of known explanations and corrections to the models will be given.

6.1. d^9 configuration

Caesium tetrachlorocuprate(II)

Cs_2CuCl_4 crystals contain squashed tetrahedral $[CuCl_4]^{2-}$ ions which approximate closely to D_{2d} symmetry [1]. In this symmetry the $^2T_2(T_d)$ term splits into $^2B_2$ and ^2E components whilst the $^2E(T_d)$ term splits into $^2B_1$ and $^2A_1$ components. Single crystal electronic spectra [2] have been assigned with the orbitally singly degenerate $^2B_2$ term lowest and ^2E about 5000 cm^{-1} above it (see Fig. 61.).

The single crystal magnetic anisotropies and average magnetic susceptibilities over the temperature range 80–300 K have been reported [3]. The results are summarized in Fig. 6.2.

Using the variation of $\bar{\mu}$ with temperature and the $^2T_2$ model summarized by Equations 5.17 and 5.21 and similar tabulations by Figgis [4] leads to an acceptable fit to the data with $v \approx -6·6$, $\lambda \approx -850$ to -550 cm^{-1} and $k \approx 1·0$ [3]. If the free ion value of $\lambda(-828$ cm$^{-1})$ is assumed this gives $\Delta \approx +5000$ cm^{-1}, an orbital singlet ground term with the orbital doublet ~ 5000 cm^{-1}

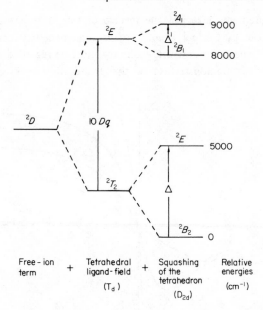

Figure 6.1. The energy diagram for the $(CuCl_4)^{2-}$ ion in $Cs_2 CuCl_4$.

above it, in excellent agreement with the electronic spectrum. However, this agreement is 'too good to be true' since the above parameters with the $^2T_2$ model predicts at 300 K that $\mu_{\parallel}$ = 1·69 and $\mu_{\perp}$ = 2·08 B.M. The magnetic

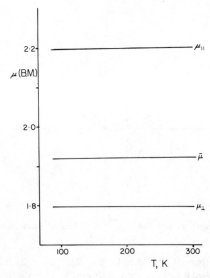

Figure 6.2. The variation with temperature of $\bar{\mu}$, $\mu_{\parallel}$, and $\mu_{\perp}$ for $Cs_2 CuCl_4$. (After Figgis, Gerloch, Lewis and Slade [3].)

anisotropy measurements give just the reverse of this, $\mu_\| = 2\cdot18$ and $\mu_\perp = 1\cdot79$ B.M. As pointed out by Gerloch *et al.* [3] 'the agreement between the powder susceptibility, the spectrum and the geometry must be regarded as fortuitous: this is again a clear demonstration of the greater usefulness of anisotropy measurements.'

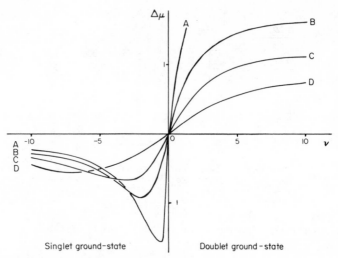

Figure 6.3. The variation of the anisotropy, $\Delta\mu = \mu_\| - \mu_\perp$, with v for the axially distorted $^2T_2$ term when $\mathbf{k} = 1\cdot0$, λ is negative, and kT/λ is (A), $-0\cdot1$; (B), $-0\cdot5$; (C), $-1\cdot0$; (D), $-2\cdot0$. (Reproduced with permission from reference 5.)

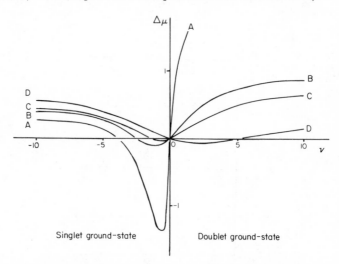

Figure 6.4. The variation of the anisotropy, $\Delta\mu = \mu_\| - \mu_\perp$ with v for the axially distorted ^2D term when $\mathbf{k} = 1\cdot0$, $\lambda = -830$ cm^{-1}, $10Dq = 7500$ cm^{-1}, $\Delta' = 100$ cm^{-1}, and kT/λ is (A), $-0\cdot1$; (B), $-0\cdot5$; (C), $-1\cdot0$; (D), $-2\cdot0$. (Reproduced with permission from reference 5.)

The reasons for this lack of agreement between theory and experiment have been discussed [3, 5]. The main conclusion is that the neglect of mixing between the components of $^2E(T_d)$ and $^2T_2(T_d)$ under spin–orbit coupling is very serious in the case of Cu(II) because of the high value of the spin–orbit coupling constant compared to the crystal field splittings. Within the $^2T_2$ model for tetrahedral copper(II) the sign of the anisotropy is uniquely determined by the sign of Δ. When Δ is positive $\mu_\parallel < \mu_\perp$ and *vice versa* (see Fig. 6.3). In the extended model where mixing with the components of $^2E(T_d)$ are included $\mu_\parallel > \mu_\perp$ occurs for a considerable range of positive values of Δ (see Fig. 6.4). In addition to the importance of the 'mixing in' of the $^2E(T_d)$ components by spin–orbit coupling it was concluded that there was a large reduction in the orbital angular momentum, $\mathbf{k} \approx 0.7$, and that this effect is anisotropic.

Tetra-(6-aminohexanoic acid) copper(II) perchlorate

This compound has been chosen as an example because it re-emphasizes the importance of including the 'mixing in' of all the excited states via spin–orbit coupling and also it has been used [6] to critically examine the approximations made in deducing the g value equation in Section 5.3.1. Single crystal electron spin resonance [6, 7] and magnetic anisotropy measurements [6] have been reported for this compound. The primary coordination geometry is very close to square planar [8] and as such may be considered as either a grossly tetragonally elongated octahedron (see 5.3.1) or a grossly squashed tetrahedron.

The observed g values are $g_\parallel = 2.301$ and $g_\perp = 2.055$ [6, 7] whilst the single crystal polarized electronic spectrum has been assigned as shown in Fig. 6.5.

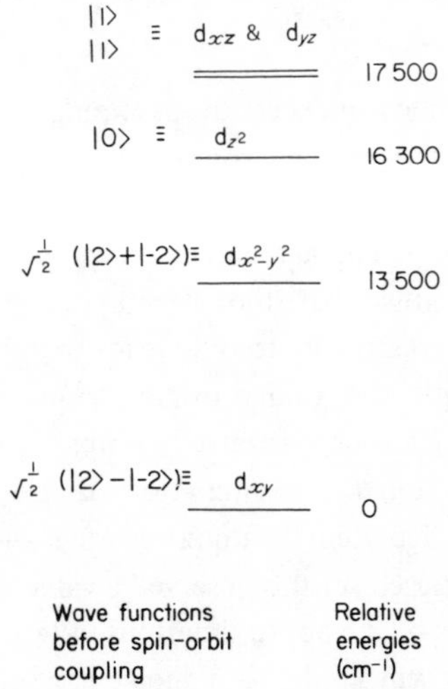

Figure 6.5. The assignment of the single crystal polarized spectrum of tetra-(6-aminohexanoic acid)copper(II)-diperchlorate.

A calculation as outlined in Section 5.3.1 using $k_\parallel \lambda \hat{L}_z \hat{S}_z$, $(k_\perp \lambda/2)[\hat{L}_x \hat{S}_x + \hat{L}_y \hat{S}_y]$, $(k_\parallel \hat{L}_z + 2 \cdot 0023\, \hat{S}_z)\beta$ and $(k_\perp \hat{L}_x + 2 \cdot 0023\, \hat{S}_x)\beta$ as the spin–orbit coupling and magnetic moment operators, and assuming that terms in $(\lambda/\Delta)^2$ are negligible leads to

$$g_\parallel = 2 \cdot 0023 - \frac{8k_\parallel^2 \lambda}{\Delta_1}, \tag{6.1}$$

$$g_\perp = 2 \cdot 0023 - \frac{2k_\perp^2 \lambda}{\Delta_2}. \tag{6.2}$$

where

$$\Delta_1 = E(d_{xy}) - E(d_{x^2-y^2})$$

and

$$\Delta_2 = E(d_{xy}) - E(d_{xz}, d_{yz}).$$

Using the experimental data and $\lambda = -828$ cm^{-1}, $k_\parallel = 0 \cdot 783$ and $k_\perp = 0 \cdot 765$.

If terms in $(\lambda/\Delta)^2$ are not neglected then the g value expressions become

$$g_\parallel = N^2 \left[2 \cdot 0023 - \frac{8k_\parallel^2 \lambda}{\Delta_1} + 2 \cdot 0023 \left(\frac{k_\parallel \lambda}{\Delta_1} \right)^2 - \frac{2 \cdot 0023}{2} \left(\frac{k_\perp \lambda}{\Delta_2} \right)^2 - k_\parallel \left(\frac{k_\perp \lambda}{\Delta_2} \right)^2 \right], \tag{6.3}$$

$$g_\perp = N^2 \left[2 \cdot 0023 - \frac{2k_\perp^2 \lambda}{\Delta_2} - 2 \cdot 0023 \left(\frac{k_\parallel \lambda}{\Delta_1} \right)^2 + \frac{2k_\parallel k_\perp^2 \lambda^2}{\Delta_1 \cdot \Delta_2} \right]. \tag{6.4}$$

We can examine the consequences of assuming that terms in $(\lambda/\Delta)^2$ are negligible by calculating the g values from Equations 6.3 and 6.4 using the values of k deduced from 6.1 and 6.2. When this is done, $N^2 = 0 \cdot 9973$, not unity, and $g_\parallel = 2 \cdot 300$ and $g_\perp = 2 \cdot 050$. The agreement for $g_\parallel$ is excellent whilst that for $g_\perp$, although good, does show that the effects of terms in $(k\lambda/\Delta)^2$ are not completely negligible. Increasing $k_\perp$ to $0 \cdot 80$ leads to $g_\parallel = 2 \cdot 300$ and $g_\perp = 2 \cdot 055$ in excellent agreement with the values originally taken. However, the relative magnitudes of $k_\parallel$ and $k_\perp$ are now reversed compared with those deduced from Equations 6.1 and 6.2. Further examination of this problem shows that the errors in the k values calculated by Equations 6.1 and 6.2 compared with 6.3 and 6.4 become larger the closer the observed g value is to $2 \cdot 0023$. Although we have discussed the effects of the approximations in terms of a crystal field model the same considerations will apply to a molecular orbital treatment, since the molecular orbital coefficients will appear in the g value equations accompanying terms in (λ/Δ) and $(\lambda/\Delta)^2$.

The above discussion is based on a simple perturbation approach which considers the effects of the crystal field and spin–orbit coupling as sequential perturbations. It is now instructive to examine the effects of considering the crystal field and spin–orbit coupling as simultaneous perturbations and solving the secular equation exactly. The results of such a calculation have been reported [6] for $\Delta_1 = 12,687$ cm^{-1}, $\Delta_2 = 17,428$ cm^{-1}, $\lambda = -828$ cm^{-1}, $k_\| = 0.78$ and $k_\perp = 0.83$. The ground state wave-functions derived from an exact solution of the secular equations are:

$$\phi_1 = 0.7411|2, \tfrac{1}{2}\rangle - 0.6709|-2, \tfrac{1}{2}\rangle - 0.02688|-1, -\tfrac{1}{2}\rangle,$$
$$\phi_2 = 0.6709|2, -\tfrac{1}{2}\rangle - 0.7411|-2, -\tfrac{1}{2}\rangle + 0.02688|1, \tfrac{1}{2}\rangle \tag{6.5}$$

giving $g_\| = 2.307$ and $g_\perp = 2.057$.

The simple perturbation approach outlined previously leads to the ground state wave-functions

$$\phi_1 = 0.7422|2, \tfrac{1}{2}\rangle - 0.6702|-2, \tfrac{1}{2}\rangle - 0.02785|-1, -\tfrac{1}{2}\rangle,$$
$$\phi_2 = 0.6702|2, \tfrac{1}{2}\rangle - 0.7422|-2, -\tfrac{1}{2}\rangle + 0.02785|1, \tfrac{1}{2}\rangle \tag{6.6}$$

and $g_\| = 2.317$, $g_\perp = 2.061$.

Although the difference in the ground state wave-functions calculated by the two methods are relatively small the effects on the calculated g values are very significant. It is interesting to note that in this particular case the most severe effect is on $g_\|$ rather than on $g_\perp$ as it was in the previous discussion. The differences in the ground state wave-functions calculated by the two methods have their origins in the fact that in the exact solution the spin–orbit mixing between the excited states (which in this example are much closer to each other than they are to the ground state) is taken into account, whereas this is not the case in the approximate treatment. The mixing between the excited states is reflected in small changes in their mixing into the ground state and a consequent change in the calculated g values.

The foregoing discussions emphasize that, for a given set of observed g values and electronic spectra, care must be taken to examine the effects of any approximations if the best values of **k** or molecular orbital coefficients are to be deduced.

Copper(II) ammonium sulphate hexahydrate

This compound is included because electron spin resonance measurements show that the g values are temperature dependent [9]. The single crystal X-ray structure has shown that the six coordination around the copper is based on an octahedron but there are three different bond lengths (see Fig. 6.6).

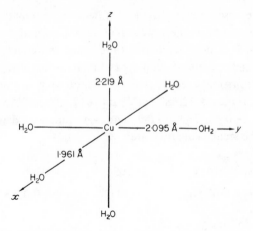

Figure 6.6. Diagram of the coordination about the copper(II) ion in copper(II) ammonium sulphate hexahydrate. (After Montgomery and Lingafelter [10].)

Single crystal electron spin resonance measurements at 20 K, which were made before the detailed crystal structure was known were interpreted with $g_z = 2\cdot40$ and $g_x = 2\cdot06$ [11]. Recent single crystal measurements on the undiluted compound at room temperature yield $g_z = 2\cdot360, g_y = 2\cdot218, g_x = 2\cdot071$ where x, y, z are specified in Fig. 6.6. At 160 K the corresponding values are $g_z = 2\cdot423, g_y = 2\cdot131$, and $g_x = 2\cdot070$ [2]. Since the g values are determined by the composition of the ground state they should be temperature independent providing that the ground state composition remains constant. Thus in the present compound the composition of the ground state must be temperature dependent which directly reflects a change in the electronic structure. This change in the electronic structure may be attributed to changes in the coordination sphere and bonding. It would be of considerable interest to determine the crystal structure at lower temperatures in order to detect any changes in the coordination sphere. Although there is always the possibility of changes in structure with temperature this example and that of $(Et_4N)_2NiCl_4$ (see 6.2) are two cases where such changes have been detected.

6.2. d⁸ configuration

Tetraethylammonium tetrachloronickelate(II), $(Et_4N)_2NiCl_4$

This compound contains essentially a tetrahedral nickel(II) complex and thus would give rise to a cubic field $^3T_1$ ground term. However, a single crystal X-ray structural analysis of $(Et_4\overset{.}{N})_2NiCl_4$ [12] has shown that the $[NiCl_4]^{2-}$ units are slightly distorted which results in a tetragonal elongation.

The average magnetic susceptibilities [13] and the anisotropies [14] of this compound have been reported over the temperature range 80–300 K. The magnetic behaviour expressed in terms of μ *vs.* T is shown in Fig. 6.7. One very significant feature revealed by the anisotropy measurements is a discontinuity in the variation of the principal susceptibilities with temperature at between 210–230 K, which is not observed in the measurement of the average susceptibility. This again demonstrates the advantages in measuring magnetic anisotropies. The abrupt change in the magnetic anisotropy has been tentatively ascribed to a change in molecular geometry [14].

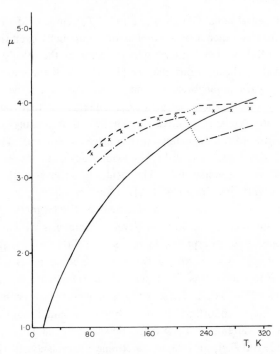

Figure 6.7. The variation of μ with temperature for $(Et_4N)_2NiCl_4$. × Experimental $\bar{\mu}$; – – – –, experimental $\mu_\perp$; – · – · –, experimental $\mu_\parallel$; ———, calculated $\bar{\mu}$ for $^3T_1$ weak crystal field term with $\lambda = -335$ cm^{-1}.

Before the magnetic anisotropy was measured the average magnetic susceptibility measurements were interpreted [13] on the basis of an axial distortion of the $^3T_1$ term described in Chapter 5. This approach was found necessary because the variation of $\bar{\mu}$ with temperature deviated considerably from that expected for a regular tetrahedron, see Fig. 6.7. Assuming that the weak crystal field approach ($A = 1·5$) was appropriate then very good agreement between theory and experiment over the temperature range 80–300 K was obtained using $k = 1·00$, $\lambda = -194$ cm^{-1}, $\Delta = 718$ cm^{-1}, see Fig. 6.7. Using a more complete model, which takes account of any crystal field and spin–orbit mixing with terms other than the $^3T_1$ ground term, and using the magnetic anisotropy above the transition point M. Gerloch and R.C. Slade [14] obtained $k = 0·67 \pm 0·02$,

$\lambda = -138 \pm 5 \, \text{cm}^{-1}$ and $\Delta = 160 \, \text{cm}^{-1}$. However, in view of the discontinuity in the magnetic anisotropy and what is probably an inappropriate ratio of the crystal field parameters Dq/Cp used in the latter calculation, the various parameters deduced from either model are of doubtful significance.

6.3. d⁷ configuration

Octahedral cobalt(II) complexes

Spin-free octahedral cobalt(II) gives rise to a $^4T_{1g}$ cubic field ground term. The average magnetic moments of a number of compounds over the temperature range 80–300 K have been interpreted in terms of the axially distorted $^4T_1$ model outlined in the previous chapter [15]. In all of the compounds studied there were always ambiguities in the parameters which would adequately account for the data. The most serious ambiguity was that the data could be interpreted with Δ either positive or negative, thus making it impossible to correlate the ground state splittings with molecular geometry. The ambiguity in interpretation could be due to either the inadequacy of the experimental data or the limitations of the model itself. The results of an investigation of these two factors have recently been reported [16].

This later work involved the measurement of the magnetic anisotropy over the temperature range 80–300 K of several complexes of known geometry. These data were then interpreted in terms of a model which allowed for any crystal field or spin–orbit mixing between all the components which arise from the ^4F and ^4P free ion terms. The results of this treatment are complex and will not be discussed in detail. These authors however draw some general conclusions, which are important. In both trigonally and tetragonally distorted situations the measurement of the magnetic anisotropy alone, over the temperature range 80–300 K, is unable to determine unambiguously the sign of the splitting of the $^4T_{1g}$ octahedral ground term. Recourse to the available electronic spectra of the compounds helps to remove some of the ambiguities in the parameterization but does not do so completely. Further, it is concluded that in the examples they studied similar splitting patterns were produced in which spin–orbit coupling effects predominate whatever the sense of the distortion.

Tetrahedral cobalt(II) complexes

Caesium pentachlorocobaltate(II) contains discrete $[CoCl_4]^{2-}$ units which have been shown, by single crystal X-ray studies, to be slightly elongated tetrahedra, the Cl–Co–Cl angle being 106° compared with 109° 28′ for a regular tetrahedron [17]. The magnetic anistropy has been determined over the temperature range 80–300 K [18] and 2–80 K [19]. E.s.r. studies at 4 and 77 K [19] give $g_{\parallel}$ = 2·41 ± 0·02, $g_{\perp}$ = 2·33 ± 0·04, and the zero-field splitting δ = 8·6 cm⁻¹. The variation of $\overline{\chi}$ with temperature reported by these two sets of workers are not in perfect agreement in that Van Stapele *et al.* find Curie law behaviour above 80 K whereas Figgis *et al.* find Curie–Weiss law behaviour with $\theta = 8°$. This difference although small has very important consequences as far as the calculation of the principal magnetic susceptibilities are concerned. These latter workers find that both $\mu_\parallel$ and $\mu_\perp$ fall with temperature rather than the behaviour predicted by Equations 5.37 and 5.42 as illustrated in Fig. 5.17. The origin of the θ value is attributed to antiferromagnetic exchange interactions between neighbouring $[CoCl_4]^{2-}$ units in the crystal lattice.

The occurrence of antiferromagnetic interactions is not supported by the more extensive measurements of Van Stapele *et al.* who find very good agreement between their results and those given by Equations 5.37, 5.42, and 5.43 after allowance is made for the T.I.P. contributions and substituting the observed *g* values for $g_\parallel = 2[1 - (4k\lambda/\Delta_1)]$ and $g_\perp = 2[1 - (4k\lambda/\Delta_2)]$.

Thus the very limited model which only considers the ground state and the cubic crystal field $^4T_2$ term seems to be sufficient to account for the observed magnetic susceptibility data. However, despite this agreement a problem still remains in that Van Stapele *et al.* report that the single crystal polarized spectrum of Cs_3CoCl_5 is consistent with $\Delta_2 < \Delta_1$ (see Fig. 5.16) whereas the e.s.r. and magnetic susceptibility results suggest the reverse.

6.4. d^6 configuration

Hexa-aquo iron(II) fluorosilicate, $Fe(H_2O)_6SiF_6$, and hexa-aquo iron(II) sulphate ammonium sulphate, $(NH_4)_2Fe(SO_4)_26H_2O$

These two compounds provide a very interesting comparison between structural and magnetic properties. Both compounds contain the essentially octahedral $[Fe(H_2O)_6]^{2+}$ ion. However, the room temperature magnetic moment of $Fe(H_2O)_6(NH_4)_2(SO_4)_2$ is 5·5 B.M. whilst that of $Fe(H_2O)_6SiF_6$ is 5·2 B.M. Along with this difference in magnetic moments the octahedron in $Fe(H_2O)_6(NH_4)_2(SO_4)_2$ is approximately tetragonally distorted [20], whilst it is trigonally distorted in $Fe(H_2O)_6SiF_6$ [21]. The principal magnetic suscepti-bilities of $Fe(H_2O)_6SiF_6$ over the temperature ranges 1·5–103 K [22] and 77–300 K [23] and the average magnetic susceptibilities over the range 80–300 K [24] have been reported. Similarly average magnetic susceptibilities [24, 25] and magnetic anisotropies [23, 26] of $Fe(H_2O)_6(NH_4)_2(SO_4)_2$, mainly over the temperature range 80–300 K have been measured by a number of authors.

The crystal structure data, the variation of $\overline{\mu}$ with temperature and the observation of magnetic anisotropy in both of these compounds demonstrates the inapplicability of the octahedral model described in Chapter 4. However, the axially distorted $^5T_{2g}$ model described in Chapter 5 should be a much better approximation. A number of authors have used essentially this model and the parameters obtained are summarized in Table 6.1. Two major features emerge from these interpretations namely:

(a) there is considerable disagreement concerning the magnitude and sign of the low symmetry crystal field splitting parameter Δ, and

(b) the values of the orbital reduction parameter k seem to be in general very much lower for $Fe(H_2O)_6SiF_6$ than for $Fe(H_2O)_6(NH_4)_2(SO_4)_2$. At first sight this difference in the orbital reduction factor between two similar compounds and the large difference in the room temperature magnetic moments seems surprising.

Table 6.1. *Parameters required to interpret the magnetic susceptibilities of* $Fe(H_2O)_6SiF_6$ *and* $Fe(H_2O)_6(NH_4)_2(SO_4)_2$ *using the axially distorted* $^5T_{2g}$ *model.*

Δ cm^{-1}	$-\lambda$ cm^{-1}	k	Reference
$Fe(H_2O)_6SiF_6$			
750	100	0·7	19
1200	100	1·0	18
950 to 1200	100	0·7–1·0	*
560 to 760	80	0·6–0·9	*
$Fe(H_2O)_6(NH_4)_2(SO_4)_2$			
410	90	1·0	19
400	100	1·0	18
-650 to -270	80	1·0	†
-1000	–	–	*

* A.S. Chakravarty and E. König (1967). *Theor. chim. Acta.*, **9**, 151. These authors suggested a temperature variation in Δ.
† A. Bose, A.S. Chakravarty, and R. Chatterjee (1961). *Proc. Roy. Soc.*, **A261**, 207. A temperature variation of Δ was invoked.

A more detailed theoretical interpretation by Gerloch *et al.* [23] in which the components of the $^5E_g(O_h)$ term were included in the calculation has satisfactorily explained these differences. Inclusion of the components of the $^5E_g(O_h)$ term in the calculation is particularly important in the case of a trigonal distortion for the following reason: in the trigonal distortion the $^5E_g(O_h)$ term does not split but transforms as ^5E_g, whilst the $^5T_{2g}(O_h)$ term splits into ^5E_g and $^5A_{1g}$ components (see Fig. 6.8).

Under these conditions the crystal field can mix the two ^5E_g terms, thus the excited term can affect the ground term components by 'mixing in' *via* both the crystal field and spin–orbit coupling. In contrast to this in the tetragonally distorted situation the components from the $^5E_g(O_h)$ term can only mix into the ground term components *via* spin–orbit coupling.

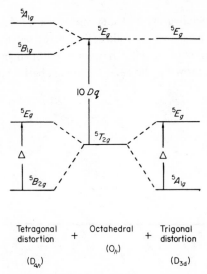

Figure 6.8. The splitting diagram for spin-free iron(II) in crystal fields of O_h, D_{4h}, *and* D_{3d} symmetry.

Although this more complete model does not lead to a unique set of parameters it leads to the conclusion that Δ is positive for both compounds. In the case of $Fe(H_2O)_6SiF_6$ it appears that the parameters **k** and Δ which fit the experimental data are not independent, e.g. when **k** = 1·0, $\Delta \approx 200$ cm^{-1}, whilst for **k** = 0·7, $\Delta \approx 750$ cm^{-1}. Despite this lack of uniqueness in the parameters some very important general conclusions can be drawn from this treatment [23].

(a) The lower value of $\bar{\mu}$ for $Fe(H_2O)_6SiF_6$ compared with $Fe(H_2O)_6$-$(NH_4)_2(SO_4)_2$ can be attributed to the trigonal distortion in the former compound. This distortion allows the excited ^5E_g term to be mixed into the ground term by both the crystal field and spin–orbit coupling. The magnetic properties are very sensitive to this mixing and in the present case it should not be ignored. Also the suggestion by Figgis *et al.* [24] that in these compounds the lowering of $\bar{\mu}$ indicates a lower value of the orbital reduction factor **k** is not necessarily correct.

(b) At least in trigonally distorted compounds the sensitivity of the magnetic properties to small distortions from regular octahedral symmetry makes the interpretation of data obtained from dilution in related host lattices a very hazardous process, particularly if the geometry determined for the pure compound is assumed to be applicable.

6.5. d^2 configuration

Ammonium vanadium alum, $(NH_4)V(SO_4)_2 12H_2O$

The coordination about the vanadium(III) ion in this compound consists of a trigonally distorted octahedron of water molecules [27], and we would not expect the magnetic behaviour to follow that predicted in Chapter 4 for the $^3T_{1g}$ term; as shown by Fig. 6.9. The model outlined in Chapter 5 for the axially distorted $^3T_{1g}$ term may however be a much better approximation.

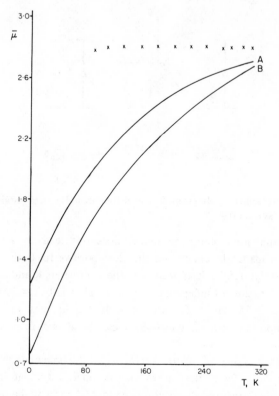

Figure 6.9. The variation of $\bar{\mu}$ with temperature for $(NH_4)V(SO_4)_2 12H_2O$. x, experimental; ———, calculated for the $^3T_{1g}$ term with $\lambda = 108$ cm^{-1}, (A) the medium crystal field and (B) the weak crystal field approximations.

The experimental data for this compound are confined to the average magnetic susceptibility over the temperature range 1–300 K [28–31]. Van der Handel and Siegert [28] made most of their measurements below 20 K, the range above this being represented by only four determinations. Information on the accuracy of the measurements and purity of the sample was not reported by these authors. Chakravarty [30] has also measured the susceptibility at four temperatures between 70 and 300 K and found good agreement with J. van der Handel and A. Siegert. J.J. Fritz and H.L. Pinch [29] measured the susceptibility between 1 and 10 K on samples which were known to have oxidized only very slightly during the course of the measurements. Similarly, B.N. Figgis, J. Lewis, and F.E. Mabbs [31] measured the susceptibility between

80–300 K on samples of high purity. In view of the purity of the samples used by these two sets of authors [29, 31], their results will be used for subsequent interpretation.

Examination of the electronic spectrum of aqueous solutions of vanadium(III) enables the mixing between the $^3T_1(F)$ and $^3T_1(P)$ terms to be determined. This leads to $A = 1\cdot2$. Using the data tabulated in Chapter 5 we find that the calculated variation of $\bar\mu$ with temperature is greater than that observed even when $v = 10$. The free ion value of λ is 105 cm^{-1} which suggests that the splitting of the $^3T_{1g}$ term by the low symmetry crystal field is at least 1000 cm^{-1}. Under these conditions it is more convenient to consider the low symmetry crystal field as the dominant perturbation followed by spin–orbit coupling. The expression for $\bar\mu$ using this approximation has been given by Siegert [32]:

$$\tfrac{1}{8}\bar\mu^2 = C(\delta/2kT)\frac{[1 - (1 - \delta/2kT)\exp(-\delta/kT)]}{[1 + 2\exp(-\delta/kT)]} + 10^{-4}T, \qquad (6.7)$$

where the Curie constant $\left.\begin{array}{l} C = 1 - 4D^2\lambda/3\Delta \\ \text{and } \delta = (D^2\lambda^2/\Delta)(1 - 2D\lambda/\Delta) \end{array}\right\}$. $\qquad (6.8)$

$D = A\mathbf{k}$ where $\mathbf{k}$ is the orbital reduction parameter in the spin–orbit coupling and magnetic moment operators $k\lambda\hat{L}.\hat{S}$ and $(k\hat{L} + 2\hat{S})\beta$, respectively. The splitting parameters Δ and δ are illustrated in Fig. 6.10.

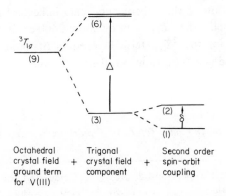

Figure 6.10. The splitting diagram for the vanadium(III) ion in a trigonally distorted octahedral crystal field.

The best experimental values for C and δ/k are 0·95 and 6·9°, respectively. The best fit between the experimental data and Equation 6.7 is given with C = 0·935 and $\delta/k = 6\cdot9°$. The agreement between theory and experiment is shown in Fig. 6.11. If we allow **k** to vary and choose Δ to give the best fit for C and δ/k simultaneously we deduce the values of Δ shown in Table 6.2.

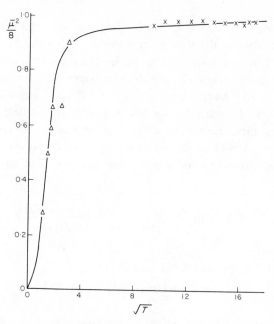

Figure 6.11. The variation of $\bar{\mu}$ with temperature for $(NH_4)V(SO_4)_2 12H_2O$. x, experimental from reference 31; Δ, experimental from reference 29; ———, calculated from Equation 6.7 with C = 0·935 and δ/k = 6·9°.

This interpretation of the experimental data indicates that the axial crystal field component splits the $^3T_{1g}$ term by about 2000 cm^{-1}, a value much larger than that found for the $^2T_{2g}$ term in caesium titanium alum which is also trigonally distorted (see 6.6). However, these conclusions should be viewed with reservation since, as in the case of caesium titanium alum, no magnetic anisotropy data are available at any temperatures and also the crystal structural

Table 6.2. *Values of the parameters in Equations 6.7 and 6.8 for ammonium vanadium alum.*

Parameter	Best experimental value	Calculated for A = 1·2 and			
		k = 1·0	0·9	0·8	0·7
C	0·95	0·935	0·935	0·935	0·935
$\delta/k°$	6·9	7·1	7·0	6·9	6·8
$10^6 \chi_{CHT}$	≯ 150	210	210	210	210
Δ cm^{-1}		3000	2400	1900	1450

Where χ_{CHT} is the temperature independent contribution to the susceptibility at high temperature.

data is not sufficiently accurate to reveal any small differences in molecular geometries which may be present in the two complexes.

6.6. d^1 configuration

Caesium titanium alum. $CsTi(SO_4)_2 12H_2O$

In this compound the titanium has a d^1 configuration giving rise to a $^2T_{2g}$ cubic field ground term. However, an X-ray structural analysis [27] shows that the geometry around the titanium(III) ion is that of a trigonally distorted octa- hedron. In view of this we would not expect the magnetic properties to conform to those derived for the $^2T_{2g}$ term in Chapter 4, and Fig. 6.12 shows that this is the case. The behaviour of the $^2T_{2g}$ term in axially symmetric fields given in Chapter 5 should be a much better approximation.

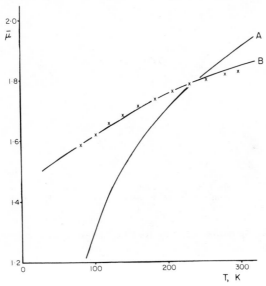

Figure 6.12. The variation of $\bar{\mu}$ with temperature for $CsTi(SO_4)_2 12H_2O$. x, experimental; ———, calculated curve (A) $^2T_{2g}$ term in O_h symmetry with λ = 145 cm^{-1}, (B) axially distorted $^2T_{2g}$ term with $v = 3.6$, $k = 0.80$, and $\lambda = 100$ cm^{-1}.

The available experimental magnetic susceptibility and e.s.r. data seem to fall into two groups. One in which the average magnetic susceptibility has been measured over the temperature range 80–300 K [33, 34], and the other in which the average magnetic susceptibility [35] and e.s.r. parameters have been determined at 4·2 K [36]. The variation of $\bar{\mu}$ with temperature over the range 80–300 K [33] is shown in Fig. 6.12. The single crystal e.s.r. study at 4·2 K gave $g_{\parallel} = 1·25$ and $g_{\perp} = 1·14$ and the low-temperature magnetic susceptibility [35] results were in agreement with these observations.

The axially distorted $^2T_{2g}$ model outlined in Section 5.2.1 can account for the variation of $\bar{\mu}$ over the range 80–300 K with $v = 3{\cdot}6, \lambda = 100$ cm^{-1} and $\mathbf{k} = 0{\cdot}80$ giving $\Delta = 360$ cm^{-1}; the agreement between the experimental and theoretical curves is shown in Fig. 6.12. These parameters result in the levels E_4 and E_6, as defined in Chapter 5, being the ground state with calculated g values of $g_{\|} = 1{\cdot}88$ and $g_{\perp} = 1{\cdot}62$. The very good agreement between the calculated and experimental values of $\bar{\mu}$ should be viewed with reservation however, for two important reasons. Firstly, there is no measurement of the anisotropy in μ which would enable us to confirm the choice of parameters and secondly the calculated g values are in very poor agreement with those observed at 4·2 K.

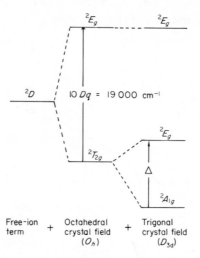

Figure 6.13. The splitting diagram for the ^2D term in a trigonally distorted octahedral crystal field.

The reasons for this lack of agreement between the magnetic susceptibilities and the g values have been investigated by Gladney and Swalen [37]. These authors included the components from the cubic field ^2E_g term in their calculation (see Fig. 6.13). This excited state can make two contributions to the magnetic susceptibility. The first of these contributions arises because the crystal field can mix the excited ^2E state with the ^2E arising from the $^2T_{2g}(O_h)$. This mixing-in of the excited wave-functions is proportional to $v'/10Dq$, where v' (using Gladney and Swalen's nomenclature) is the value of the matrix element between the two E states under the trigonal crystal field. The second contribution will arise from the mixing between the components from the ^2E excited states and the lower lying states under the action of spin–orbit coupling. This 'mixing-in' is proportional to $\lambda/10Dq$ and would be expected to be smaller than the crystal field mixing, but not necessarily unimportant. Using this more complete model and allowing for anisotropy in the orbital reduction parameter, **k**, enabled exact agreement with the observed g values to be obtained but not with a unique set of parameters (some examples of the parameters produced by Gladney and Swalen [37] are shown in Table 6.3). However, it was reported [37] that these parameters do not reproduce the variation of $\bar{\mu}$ in the range 80–300 K at all well.

Table 6.3. *Parameters which fit the g values for* $CsTi(SO_4)_2 12H_2O$.

Trigonal field parameters		Orbital reduction parameters	
$\Delta\,(cm^{-1})$	$v'\,(cm^{-1})$	$k_{\parallel}$	$k_{\perp}$
500	−2800	0·84	0·81
250	−1500	1·00	0·72
170	0	0·90	0·63
250	1100	1·00	0·54
500	2400	0·85	0·44

The overall situation on $CsTi(SO_4)_2 12H_2O$ is at the moment unsatisfactory in that even the more complete model does not account for all the available data. Where data can be accounted for, this model does not lead to a unique set of parameters. Clearly more detailed and accurate anisotropy data is required over the whole temperature range 300–4 K and equally importantly we need to know whether or not there are any significant structural changes over this temperature range. The magnetic properties may well be very sensitive to even small structural changes and perhaps it is too optimistic to hope that one unique set of parameters will account for the magnetic data over such a wide temperature range.

Concluding remarks

In this chapter we have taken a small number of compounds of known structure and examined critically the interpretation of the magnetic susceptibility and e.s.r. data based on the theoretical models discussed in the previous chapters. These models are normally quite good at accounting for the major features observed in the data. However, as more detailed data become available it is evident that more sophisticated models than those discussed in Chapters 4 and 5 will be required. Also, because the models assume that the molecular geometry is independent of temperature, it is often unclear whether deviations from a particular model are due to changes in geometry or to other limitations in the model. In two of the examples discussed the magnetic properties strongly suggest that the molecular structures change with temperature, although this has yet to be confirmed by single crystal X-ray structural determinations. However, it should be quite clear that, as the magnetic susceptibility and e.s.r. properties of transition metal complexes depend upon the wave-functions describing the thermally populated energy levels, these types of measurements can make a significant contribution to our understanding of the electronic structures of transition metal complexes.

REFERENCES

1. L. Helmholz and R.F. Kruh (1952). *J. Amer. Chem. Soc.,* **76**, 1176; B. Morosin and E.C. Lingafelter (1961). *J. Phys. Chem.,* **65**, 50.

2. J. Ferguson (1964). *J. Chem. Phys.,* **40**, 3406.

3. B.N. Figgis, M. Gerloch, J. Lewis and R.C. Slade (1968). *J. Chem. Soc. (A),* 2028.

4. B.N. Figgis (1961). *Trans. Faraday Soc.,* **57**, 204.

5. M. Gerloch (1968). *J. Chem. Soc. (A),* 2023.

6. C.D. Garner, P. Lambert, F.E. Mabbs and J.K. Porter (1972). *J. Chem. Soc. Dalton,* 320.

7. R.J. Dudley, B.J. Hathaway and P.G. Hodgson (1970). *J. Chem. Soc. (A),* 3355.

8. R. Österberg, B. Sjöberg and R. Söderquist (1970). *Chem. Comm.,* 1408.

9. F.E. Mabbs, C.D. Garner and J.K. Porter (1972), unpublished results.

10. H. Montgomery and E.C. Lingafelter (1966). *Acta Cryst.,* **20**, 659.

11. B. Bleaney, R.P. Penrose and B.I. Plumpton (1949). *Proc. Roy. Soc. (London),* **A198**, 406; B. Bleaney, K.D. Bowers and B.I. Plumpton (1955). *Proc. Roy. Soc. (London),* **A228**, pp. 147, 157, 166; B. Bleaney, K.D. Bowers and D.J.E. Ingram (1951). *Proc. Phys. Soc.,* **64**, 758.

12. G.D. Stucky, J.B. Folkers and T.J. Kistenmaker (1967). *Acta. Cryst.,* **23**, 1064.

13. B.N. Figgis, J. Lewis, F.E. Mabbs and G.A. Webb (1966). *J. Chem. Soc. (A),* 1411.

14. M. Gerloch and R.C. Slade (1969). *J. Chem. Soc. (A),* 1022.

15. B.N. Figgis, M. Gerloch, J. Lewis, F.E. Mabbs and G.A. Webb (1968). *J. Chem. Soc. (A),* 2086.

16. M. Gerloch and P.N. Quested (1971). *J. Chem. Soc. (A),* 3729; M. Gerloch, P.N. Quested and R.C. Slade (1971). *J. Chem. Soc. (A),* 3741.

17. B.N. Figgis, M. Gerloch and R. Mason (1964). *Acta. Cryst.,* **17**, 506.

18. B.N. Figgis, M. Gerloch and R. Mason (1964). *Proc. Roy. Soc. (London),* **A279**, 210.

19. R.P. van Stapele, H.G. Beljers, P.F. Bongers and H. Zijlstra (1966). *J. Chem. Phys.* **44**, 3719.

20. H. Montgomery, R.V. Chastain, J.J. Natt, A.M. Witkouska and E.C. Lingafelter (1967). *Acta Cryst.*, **22**, 775.

21. W.C. Hamilton (1962). *Acta Cryst.*, **15**, 353.

22. L.C. Jackson (1929). *Phil. Mag.*, **4**, 269; T. Ohtsuka (1959). *J. Phys. Soc. Japan*, **14**, 1245.

23. M. Gerloch, J. Lewis, G.G. Phillips and P.N. Quested (1970). *J. Chem. Soc. (A)*, 1941.

24. B.N. Figgis, J. Lewis, F.E. Mabbs and G.A. Webb (1967). *J. Chem. Soc. (A)*, 442.

25. G. Foex (1921). *Ann. Phys.*, **16**, 174; L.C. Jackson (1924). *Phil. Trans.*, **A224**, 1.

26. K.S. Krishnan, N.C. Chakravorty and S. Banerjee (1933). *Phil. Trans.*, **A232**, 99; S. Datta (1954). *Ind. J. Phys.*, **28**, 239; A. Bose (1948). *Ind. J. Phys.*, **31**, pp. 74, 483.

27. W. Lipson (1935). *Proc. Roy. Soc. (London)*, **A151**, 347.

28. J. van der Handel and A. Siegert (1937). *Physica*, **4**, 871.

29. J.J. Fritz and H.L. Pinch (1956). *J. Amer. Chem. Soc.*, **78**, 6223.

30. A.S. Chakravarty (1958). *Ind. J. Phys.*, **32**, 447.

31. B.N. Figgis, J. Lewis and F.E. Mabbs (1960). *J. Chem. Soc.*, 2480.

32. A. Siegert (1937). *Physica*, **4**, 138.

33. B.N. Figgis, J. Lewis and F.E. Mabbs (1963). *J. Chem. Soc.*, 2473.

34. A. Bose, A.S. Chakravarty and R. Chatterjee (1960). *Proc. Roy. Soc. (London)*, **A255**, 145.

35. R.J. Benzie and A.H. Cooke (1951). *Proc. Roy. Soc. (London)*, **A209**, 269.

36. B. Bleaney, G.S. Bogle, A.H. Cooke, R.J. Duffus, M.C.M. O'Brien and K.W.H. Stevens (1955). *Proc. Phys. Soc. (London)*, **A68**, 57.

37. H.M. Gladney and J.D. Swalen (1965). *J. Chem. Phys.*, **42**, 1999.

7 | The magnetic properties of polynuclear transition metal complexes

7.1. Introduction

The previous chapters have dealt with the magnetic behaviour of magnetically dilute systems. There is also a large number of systems such as oxides and halides which are said to be magnetically concentrated, where the electron spins on adjacent paramagnetic centres are strongly coupled to each other. This coupling between the electronic spins leads to antiferromagnetism when the spins are aligned anti-parallel and to ferromagnetism when the spins are aligned parallel. However, intermediate between magnetically concentrated and magnetically dilute systems there is a large range of polynuclear transition metal complexes in which exchange interactions take place over a small number of paramagnetic centres. A simple discussion of the magnetic properties of this latter class of compounds forms the major part of this chapter. We will, however, confine ourselves to the treatment of *antiferromagnetic* interactions between *adjacent* paramagnetic ions in these complexes, a simplification which the reader should remember when applying any of the conclusions to actual systems.

Broadly speaking antiferromagnetic systems may be divided into two main types – those in which the magnetic exchange occurs between centres in the same molecule (intramolecular antiferromagnets), and those in which the exchange interaction extends over a large number of centres throughout a crystal lattice (intermolecular antiferromagnets). Polynuclear transition metal complexes fall into the first group whilst the latter group is typified by 'simple' halide and oxide systems.

An intermolecular antiferromagnetic system normally exhibits the characteristic variation of magnetic susceptibility with temperature shown in Fig. 7.1.

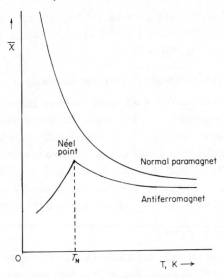

Figure 7.1. A comparison of the characteristic variation of susceptibility with temperature for normal paramagnetic and antiferromagnetic materials.

Starting at high temperatures, the magnetic susceptibility increases as the temperature is lowered, but more slowly than expected for a magnetically dilute system. This variation of the susceptibility with temperature often leads to a Curie–Weiss law behaviour with appreciable positive values of θ, and hence the dangerous assumption often made that large values of θ are associated with the presence of antiferromagnetic interactions. Some of the dangers in this assumption were discussed in Chapter 4. However, if the temperature is lowered sufficiently, a fairly sharp maximum in the magnetic susceptibility is reached at a temperature, T_N (the Néel point), below which the susceptibility decreases rapidly with temperature. A simple interpretation of this magnetic behaviour is that above T_N the thermal energy available to the system is sufficient to overcome the aligning forces resulting in the magnetic behaviour resembling that of a magnetically dilute system. At T_N, these two effects balance each other, hence the maximum in the susceptibility, whilst below this critical temperature the forces aligning the spins anti-parallel predominate, and the magnetic susceptibility falls. There will normally be a preferred direction for the orientation of the electronic spins throughout the lattice, below T_N the magnetic susceptibility along this direction will be dependent on the strength of the applied magnetic field. The magnetic susceptibility perpendicular to this direction will be of the normal field independent type. Thus below the Néel point the magnetic susceptibility of a powdered specimen may show some dependence on magnetic field. Intramolecular antiferromagnetic systems exhibit many of the general features of intermolecular antiferromagnetics, the major differences being that the maximum in the susceptibility is normally much broader, and the magnetic susceptibilities are not usually dependent on the strength of the applied magnetic field.

7.2. Mechanisms of antiferromagnetic exchange interactions

Having concluded that antiferromagnetism involves the interaction between electronic spins on neighbouring metal atoms, the most obvious question which arises is, 'What is the mechanism of this interaction?' It is generally accepted that the mechanism of the exchange interaction involves the mutual pairing of electronic spins *via* some form of orbital overlap, analogous to the formation of a chemical bond. The following two mechanisms are usually used to account for antiferromagnetic exchange: (a) direct interaction, and (b) superexchange.

7.2.1. DIRECT INTERACTION

As the term implies, this mechanism involves direct overlap between the orbitals containing the unpaired electrons, leading to mutual pairing in the ground state. This mechanism is the one considered to be responsible for exchange interaction in copper(II) acetate monohydrate. In this molecule weak overlap between the $d_{x^2-y^2}$ orbitals on each copper atom, giving a δ-bond has been proposed [1], see Fig. 7.2. This weak overlap leads to a diamagnetic spin-singlet ground state for the molecule.

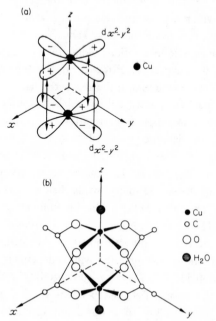

Figure. 7.2. (a) An illustration of the δ-bondong in and (b) the structure of copper(II) acetate monohydrate. (After Figgis and Martin [1].)

As well as this spin-singlet ground state the interaction also gives rise to an excited paramagnetic spin-triplet state. The observed magnetic behaviour, Fig. 7.3, which will be discussed in more detail later, arises from thermal population of the spin-triplet state. The overlap between the metal $d_{x^2-y^2}$ orbitals is relatively weak and it is not considered to be responsible for holding the molecule together, and in the normal sense it would not be considered to constitute a metal–metal bond.

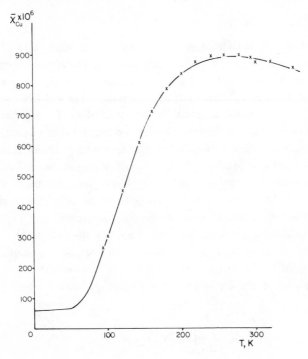

Figure 7.3. The variation of $\bar{\chi}_{Cu}$ with T for copper(II) acetate monohydrate. x, experimental; ———, calculated from Equation 7.10 with $\bar{g} = 2 \cdot 16$, $J = -295$ cm^{-1}, and $N\alpha = 60 \times 10^{-6}$ c.g.s.

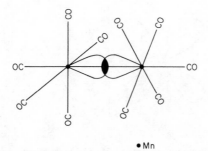

Figure 7.4. An illustration of the metal–metal bonding in $Mn_2(CO)_{10}$.

More extreme examples of the direct interaction involving transition metal d-orbitals occur in many carbonyl and halide cluster compounds. For example in $(CO)_5Mn-Mn(CO)_5$ there is considered to be a strong overlap between the manganese d-orbitals, which formally may be expected to contain an unpaired electron in the monomer (see Fig. 7.4). This overlap leads to a metal–metal bond which is responsible for holding the molecule together and pairs the odd electrons leading to a diamagnetic ground state for the molecule.

7.2.2. SUPEREXCHANGE

This mechanism for antiferromagnetism involves the interaction of electrons with opposite spins on the two interacting ions *via* an intermediate diamagnetic anion. This mechanism is usually the one postulated to explain the antiferromagnetism of oxide and fluoride lattice compounds, and was first proposed by Kramers [2] to explain the magnetic behaviour of MnO. The mechanism again involves orbital overlap, but instead of only the metal d-orbitals being involved, the participation of filled orbitals on the intervening anion must also be considered. In order to illustrate this mechanism more fully we will consider the hypothetical oxide system M_2O, in which M is a transition metal ion with a single d-electron. In a linear M–O–M arrangement the interaction may occur in two ways; either via a σ-bonding or a π-bonding mechanism (see Fig. 7.5). The σ-bonding mechanism is represented, Fig. 7.5(a), by the overlap of d_{z^2} orbitals on the metal ions with a p_z orbital on the oxygen. A simple pictorial representation of the exchange process involves an electron with spin 'up' on M_1 pairing with one of the electrons in the oxygen p_z orbital which has a spin

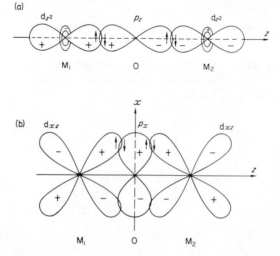

Figure 7.5. Superexchange in a linear M–O–M system. An example of superexchange *via* (a) σ-bonding, and (b) π-bonding.

'down'. This leaves the other electron in the oxygen p_z orbital with a spin 'up'. If this electron interacts with the electron on M_2 in an antiferromagnetic sense then the electron in the d_{z^2} orbital on M_2 must have its spin 'down'. Similar considerations can be applied to the superexchange mechanism *via* π-bonding (Fig. 7.5(b)). These rather crude pictorial representations enable us to envisage how the spins on the interacting metal ions may be aligned in opposite directions and how they may become effectively paired leading to a minimum number of unpaired spins in the ground state of the system.

The superexchange mechanism just described may be extended to systems in which more than one anion intervenes between the paramagnetic ions. This type of mechanism has been proposed to explain the antiferromagnetic interaction which occurs in certain salts of the $IrCl_6^{2-}$ anion [3]. The exchange between nearest neighbouring iridium atoms occurs via two intervening chloride ions in a system of the form

$$
\begin{array}{ccc}
 & \text{Cl----Cl} & \\
\text{Ir} & & \text{Ir} \\
 & \text{Cl----Cl} & \\
\end{array}
$$

7.3. Antiferromagnetism in Polynuclear transition metal complexes

The interpretation of the magnetic susceptibilities of polynuclear transition metal complexes involves two complementary approaches. These two approaches are molecular orbital theory and the dipolar coupling approach of Van Vleck [4]. Molecular orbital theory takes into account the mechanism of the exchange interaction and bonding in the complex. However, it is not normally possible to readily predict or rationalize anything more than the spin multiplicity of the ground state of the molecule. On the other hand the dipolar coupling approach merely assumes the presence of an exchange interaction without reference to the mechanism, but it has the advantage that it is easy to calculate the variation of the magnetic susceptibility with temperature. A more detailed description of the use of the dipolar coupling approach will be given later in this chapter. However, before doing this a few examples of the application of molecular orbital theory to polynuclear systems will be given.

The example of copper(II) acetate monohydrate has already been used to illustrate one of the mechanisms of antiferromagnetic exchange interactions, and will now be used to show the application of molecular orbital theory. In the copper(II) acetate monohydrate dimer, the unpaired electron of each copper atom is considered to be in the $d_{x^2-y^2}$ orbital and the interaction to be due to direct overlap between these orbitals. The symmetry of the molecule approximates to D_{4h} and under this symmetry the $d_{x^2-y^2}$ orbitals on each copper combine to give a b_{1g} bonding and a b_{2u} anti-bonding orbital. There are two electrons to be placed in this system, and placing them both in a b_{1g} bonding orbital will result in a diamagnetic ground state for the molecule. An excited triplet state, which must be appreciably populated at normal temperatures in order to account for the observed magnetic behaviour, can be obtained by promoting an electron from the b_{1g} bonding orbital to the b_{2u} antibonding orbital, giving the configuration $(b_{1g})^1(b_{2u})^1$.

The diamagnetism of $K_4[Ru_2OCl_{10}]H_2O$, in which each ruthenium has a d^4 configuration, has also been successfully explained on a molecular orbital basis [5]. The anion has a linear Ru—O—Ru linkage and belongs to the point group D_{4h}. If the σ-bonding structure is neglected there are available for π-bonding six t_{2g} orbitals from the pair of approximately octahedrally co-ordinated rutheniums, and two oxygen p_π orbitals. In D_{4h} symmetry these orbitals combine to give bonding orbitals E_u^b, non-bonding orbitals E_g^{nb}, B_{2g}^{nb}, and B_{1u}^{nb}, and antibonding orbitals E_u^a. The twelve electrons in the system (four from each ruthenium and four from the two oxygen p_π-orbitals) are fed into these molecular orbitals to give the configuration: $(E_u^b)^4[(B_{2g}^{nb})^2(B_{1u}^{nb})^2(E_g^{nb})^4](E_u^a)^0$. All the electrons are paired in this scheme, compatible with the observed diamagnetism of the compound.

One example of the difficulties which can arise in the simple molecular orbital approach is illustrated by the complex cation $[(NH_3)_5Cr-O-Cr(NH_3)_5]^{4+}$, which has the same symmetry as the $[Ru_2OCl_{10}]^{4-}$ anion if a linear Cr—O—Cr bridge is assumed. In the case of the chromium complex there are ten electrons in the π-bonding system and, if we assume the same relative ordering of the molecular orbitals as in the ruthenium case, we have the configuration $(E_u^b)^4[(B_{2g}^{nb})^2(B_{1u}^{nb})^2(E_g^{nb})^2](E_u^a)^0$. If this were the ground state of the molecule then this configuration predicts that there should be two unpaired electrons per molecule, whereas magnetic susceptibility measurements are compatible with a diamagnetic ground state. A diamagnetic configuration may be obtained from the molecular orbital picture if a different ordering of the orbitals is assumed, e.g. $(E_u^b)^4[(E_g^{nb})^4(B_{2g}^{nb})^2(B_{1u}^{nb})^0](E_u^a)^0$. However, the simple molecular orbital approach cannot readily predict which configuration is the ground state or the relative separations between the other possible configurations. It will be shown later that the dipolar coupling approach correctly predicts diamagnetic ground states for the three examples discussed above providing an antiferromagnetic interaction is assumed. However this statement is not to be taken to imply that the dipolar coupling approach is not without its difficulties.

Dipolar coupling approach to the magnetic susceptibilities of polynuclear transition metal complexes

The spin–spin interaction can be treated quantitatively using the approach outlined by Van Vleck [4]. In his treatment Van Vleck notes that 'the exchange effect, though entirely orbital in nature, is, because of the exclusion principle, very sensitive to the way the spin is aligned, and is *formally equivalent* to cosine coupling between the spin magnets of the various atoms'. The cosine dependence of the coupling between the electronic spins is the same as that found in one term of the potential energy of two dipoles, hence the somewhat misleading term 'Dipolar Coupling' which has been given to this method of treating the exchange interaction.

The discussion of the application of the 'Dipolar Coupling' approach to spin–spin exchange will be confined to systems in which there is, to a first approximation, no orbital angular momentum associated with ground states of the interacting metal ions. In this approximation the magnetic exchange between nearest neighbours will be isotropic and may be represented in a spin Hamiltonian for the system by a term:

$$\mathcal{H} = -2 \sum_{\text{neighbours}} J_{ij} \hat{S}_i . \hat{S}_j \qquad (7.1)$$

where J_{ij} is an exchange integral between centres i and j. J_{ij} is negative for an antiferromagnetic interaction and positive for a ferromagnetic interaction.

If we consider the case where all the interacting ions in the system are equivalent, then the interaction can be considered to give rise to a set of spin levels for the system, each of which is specified by a new quantum number S', which can have the following values, $S' = nS, nS - 1, nS - 2, \ldots, 0$ or $\frac{1}{2}$, depending on whether nS is integral or half-integral. In this particular situation of entirely equivalent interacting centres Van Vleck has derived the following expression for the energy of each of the spin states specified by S':

$$W(S') = \frac{zJ}{n-1}[S'(S'+1) - nS(S+1)] \qquad (7.2)$$

where z = the number of nearest neighbours,

   n = the number of interacting paramagnetic ions,

   J = the exchange integral between neighbouring pairs of paramagnetic ions.

It is possible that a given value of S' may occur a number of times because of the different ways of combining the spins on the various centres. The number of times a given value of S' occurs $\Omega(S')$ (i.e. its degeneracy) is calculated from the expression:

$$\Omega(S') = \omega(S') - \omega(S' + 1) \qquad (7.3)$$

where $\omega(S')$ is the coefficient of S' in the expansion of

$$(x^S + x^{S-1} + \ldots + x^{-S+1} + x^{-S})^n \qquad (7.4)$$

In a magnetic field each of the S' levels will be split into $2S' + 1$ components which are separated by $g\beta H$. Since we are dealing with antiferromagnetic systems the smallest values of S' lie lowest and it is normally permissible to use Van Vleck's susceptibility equation since the only levels which are likely to be thermally populated will satisfy the assumption that $W_i^{(1)}H$ and $W_i^{(2)}H^2 \ll kT$ (see Chapter 1). The application of Van Vleck's susceptibility equation gives the *molar* susceptibility as

$$\bar{\chi}_M = \frac{N\beta^2\bar{g}^2}{3kT} \frac{\sum\limits_{S'} S'(S' + 1)(2S' + 1)\Omega(S') \exp(-W(S')/kT)}{\sum\limits_{S'} (2S' + 1)\Omega(S') \exp(-W(S')/kT)} \tag{7.5}$$

However, some caution in applying Van Vleck's equation to systems containing a large number of weakly interacting centres may be needed, since levels with large values of S', for which $W_i^{(1)}H$ may not be less than kT, may be thermally populated at the temperatures used to measure the magnetic susceptibilities.

When the system under consideration contains inequivalent interacting centres then the energies of the various spin states are not given by Equation 7.2, although sometimes this expression has been used with an average value for z and J. In these circumstances Kambe's [6] modification of Van Vleck's approach is more appropriate. The details of Kambe's approach will be deferred until Section 7.3.2.

The application of the dipolar coupling treatment to some simple systems will now be given in detail, and the results for other commonly occurring polynuclear arrangements will be quoted.

7.3.1. BINUCLEAR COMPLEXES

The part of the Hamiltonian representing the exchange interaction will be

$$\mathcal{H} = -2J\hat{S}_1 . \hat{S}_2 \tag{7.6}$$

Assuming each interacting ion to be identical with spin S, the possible new spin quantum numbers for the dimer are:

$$S' = 2S, 2S - 1, \ldots . 0 \tag{7.7}$$

Using Equation 7.2. the energy of each of these states is given by:

$$W(S') = -J[S'(S' + 1) - 2S(S + 1)] \tag{7.8}$$

For an antiferromagnetic interaction J is negative, and the state with the smallest value of S' lies lowest. The energies of the other spin states of the dimer relative to the lowest one are illustrated in Fig. 7.6 for various values of S for the interacting ions.

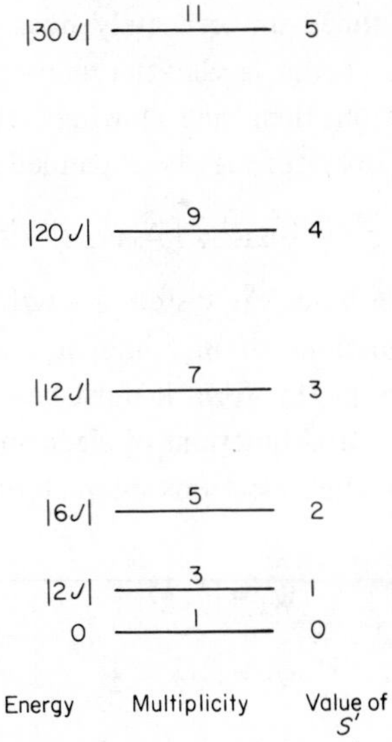

Figure 7.6. The relative energies and multiplicities of the possible spin state of a binuclear complex.

The simplest example occurs when each interacting ion has formally one unpaired electron and therefore $S = \frac{1}{2}$. There are two possible values of S', i.e. 1 and 0, with the spin-triplet ($S' = 1$) at $-2J$ above the singlet level. Before we can calculate the magnetic susceptibility from Equation 7.5 we need to know the number of times each value of S' can occur. In this particular case the question is almost trivial but the computation will be given to show the use of Equations 7.3 and 7.4. Firstly we must expand $(x^S + x^{S-1} + \ldots + x^{-S})^n$, i.e. $(x^{\frac{1}{2}} + x^{-\frac{1}{2}})^2$
$= x^1 + 2x^0 + x^{-1}$

$$\therefore \quad \Omega(S' = 1) = \omega(1) - \omega(2) = 1 - 0 = 1$$

and

$$\Omega(S' = 0) = \omega(0) - \omega(1) = 2 - 1 = 1$$

Substitution in Equation 7.5 gives:

$$\bar{\chi}_M = \frac{N\beta^2 \bar{g}^2}{3kT} \left[\frac{(1)(2)(3)(1) \exp(+2J/kT) + (0)(1)(1)(1) \exp(0)}{(3)(1) \exp(+2J/kT) + (1)(1) \exp(0)} \right]$$

$$= \frac{N\beta^2 \bar{g}^2}{3kT} \left[\frac{6 \exp(+2J/kT)}{3 \exp(+2J/kT) + 1} \right], \quad (7.9)$$

or

$$\bar{\chi}_A = \frac{\chi_M}{2} = \frac{N\beta^2 \bar{g}^2}{3kT} \left[\frac{3 \exp(2x)}{3 \exp(2x) + 1} \right], \quad (7.10)$$

where $x = J/kT$.

The treatment just outlined unfortunately does not enable us to find the wave-functions within the energy levels after spin–spin coupling. If we require the form of the wave-functions then the following method must be adopted.

The spin–spin coupling operator may be expanded as

$$-2J\hat{S}_1 \cdot \hat{S}_2 = -2J(\hat{S}_{z1}\hat{S}_{z2} + \hat{S}_{x1}\hat{S}_{x2} + \hat{S}_{y1}\hat{S}_{y2}). \tag{7.11}$$

The wave-functions for the binuclear system are written in the form $|m_{s1}, m_{s2}\rangle$, where all possible combinations of m_{s1} and m_{s2} must be accounted for. In writing the wave-functions in this form it must be remembered, however, that we are assuming that the orbital functions of electrons 1 and 2 are identical. For a system in which two unpaired electrons interact with each other we have the wave-functions:

$$\psi_1 = |\tfrac{1}{2}, \tfrac{1}{2}\rangle$$

$$\psi_2 = |-\tfrac{1}{2}, -\tfrac{1}{2}\rangle$$

$$\psi_3 = |\tfrac{1}{2}, -\tfrac{1}{2}\rangle$$

$$\psi_4 = |-\tfrac{1}{2}, \tfrac{1}{2}\rangle.$$

The energy of the system under the action of the perturbation represented by equation 7.11 and also the zeroth order wave-functions may be obtained by solving the appropriate secular equations.

Since we have so far not met operators of the form of (7.11) the evaluation of some of the matrix elements required in the secular determinant will be given, e.g.

$$\langle\psi_1|-2J\hat{S}_1 \cdot \hat{S}_2|\psi_1\rangle = \langle\tfrac{1}{2}, \tfrac{1}{2}|-2J(\hat{S}_{z1}\hat{S}_{z2} + \hat{S}_{x1}\hat{S}_{x2} + \hat{S}_{y1}\hat{S}_{y2}|\tfrac{1}{2}, \tfrac{1}{2}\rangle$$

$$= -2J[\langle\tfrac{1}{2}, \tfrac{1}{2}|\hat{S}_{z1}\hat{S}_{z2}|\tfrac{1}{2}, \tfrac{1}{2}\rangle + \langle\tfrac{1}{2}, \tfrac{1}{2}|\hat{S}_{x1}\hat{S}_{x2}|\tfrac{1}{2}, \tfrac{1}{2}\rangle$$

$$+ \langle\tfrac{1}{2}, \tfrac{1}{2}|\hat{S}_{y1}\hat{S}_{y2}|\tfrac{1}{2}, \tfrac{1}{2}\rangle]$$

The subscripts on the operators tell us that the particular operator only operates on the part of the wave-function describing that particular electron, e.g. S_{z1} only operates on the part of the wave-function specified by m_{s1} and leaves the remainder unchanged. Hence using this definition and the following rules

$$\hat{S}_z|\pm\tfrac{1}{2}\rangle = \pm\tfrac{1}{2}|\pm\tfrac{1}{2}\rangle$$

$$\hat{S}_x|\pm\tfrac{1}{2}\rangle = \tfrac{1}{2}|\mp\tfrac{1}{2}\rangle$$

$$\hat{S}_y|\pm\tfrac{1}{2}\rangle = \pm\frac{i}{2}\left|\mp\tfrac{1}{2}\right\rangle,$$

we have that

$$\langle \tfrac{1}{2}, \tfrac{1}{2} | \hat{S}_{z1} \hat{S}_{z2} | \tfrac{1}{2}, \tfrac{1}{2} \rangle = \tfrac{1}{2} \cdot \tfrac{1}{2} \langle \tfrac{1}{2}, \tfrac{1}{2} | \tfrac{1}{2}, \tfrac{1}{2} \rangle = \tfrac{1}{4}$$

$$\langle \tfrac{1}{2}, \tfrac{1}{2} | \hat{S}_{x1} \hat{S}_{x2} | \tfrac{1}{2}, \tfrac{1}{2} \rangle = \tfrac{1}{2} \cdot \tfrac{1}{2} \langle \tfrac{1}{2}, \tfrac{1}{2} | -\tfrac{1}{2}, -\tfrac{1}{2} \rangle = 0$$

$$\langle \tfrac{1}{2}, \tfrac{1}{2} | \hat{S}_{y1} \hat{S}_{y2} | \tfrac{1}{2}, \tfrac{1}{2} \rangle = \frac{i}{2} \cdot \frac{i}{2} \langle \tfrac{1}{2}, \tfrac{1}{2} | -\tfrac{1}{2}, -\tfrac{1}{2} \rangle = 0$$

$$\langle \psi_1 | -2J \hat{S}_1 \cdot \hat{S}_2 | \psi_1 \rangle = -J/2.$$

Evaluation of all the remaining matrix elements in the same way gives the following secular determinant

	$|\psi_1\rangle$	$|\psi_2\rangle$	$|\psi_3\rangle$	$|\psi_4\rangle$	
$\langle\psi_1|$	$\dfrac{-J}{2} - E$	0	0	0	
$\langle\psi_2|$	0	$\dfrac{-J}{2} - E$	0	0	$= 0.$ (7.12)
$\langle\psi_3|$	0	0	$\dfrac{J}{2} - E$	$-J$	
$\langle\psi_4|$	0	0	$-J$	$\dfrac{J}{2} - E$	

The solutions of this determinant give $E_1 = E_2 = E_3 = -J/2$ and $E_4 = 3J/2$, i.e. precisely the energies given by Equation 7.8. However, in the treatment we are also able to determine the wave-functions within our new energy states. This is done just as in Chapter 4, with the result:

$$\left. \begin{aligned} \phi_1 &= |\psi_1\rangle \\ \phi_2 &= |\psi_2\rangle \\ \phi_3 &= \sqrt{\tfrac{1}{2}}(|\psi_3\rangle + |\psi_4\rangle) \end{aligned} \right\} \text{ at energy } -J/2,$$

$$\phi_4 = \sqrt{\tfrac{1}{2}}(|\psi_3\rangle - |\psi_4\rangle) \text{ at energy } 3J/2.$$

These wave-functions may now be used to calculate the energies in a magnetic field by using the perturbing operator

$$\mathscr{H}' = (\hat{l}_{z1} + 2\hat{s}_{z1})\beta H + (\hat{l}_{z2} + 2\hat{s}_{z2})\beta H. \qquad (7.13)$$

N.B. The z components only are used since we make the assumption that the system is isotropic. Making the further assumption that there is no orbital angular momentum associated with our wave-functions $\mathscr{H}'$ may be rewritten as

$$\mathscr{H}' = 2\hat{S}_{z1}\beta H + 2\hat{S}_{z2}\beta H.$$

Using this operator we find the following first- and second-order Zeeman coefficients

At E = −J/2

$$W_1^{(1)} = 2\beta, \qquad W_2^{(1)} = -2\beta, \qquad W_3^{(1)} = 0$$
$$W_1^{(2)} = W_2^{(2)} = W_3^{(2)} = 0$$

At E = 3J/2

$$W_4^{(1)} = 0, \qquad W_4^{(2)} = 0$$

$$\therefore \quad \bar{\chi}_M = \frac{N\beta^2}{3kT} 3 \left[\frac{8 \exp (J/2kT)}{3 \exp (J/2kT) + \exp (-3J/2kT)} \right]$$

$$= \frac{N\beta^2}{3kT} 4 \left[\frac{6 \exp (2x)}{3 \exp (2x) + 1} \right], \tag{7.14}$$

where $x = J/kT$.

Equation 7.14 is exactly the same as 7.9 when $\bar{g} = 2$, i.e. when there is no orbital contribution.

Binuclear copper(II) complexes

Perhaps the best known and most widely studied binuclear copper(II) complex is copper(II) acetate monohydrate, the structure of which is illustrated in Fig. 7.2. The variation of the magnetic susceptibility with temperature for each copper ion in the molecule is shown in Fig. 7.3 and is compared with that given by Equation 7.10 for $\bar{g} = 2 \cdot 16$ [7], $J = -295$ cm^{-1}, and a temperature independent contribution, $N\alpha = 60 \times 10^{-6}$ c.g.s. It is important to briefly discuss the values of $\bar{g}$ and $N\alpha$ used to account for the magnetic susceptibility. If, as initially assumed, each interacting ion has no orbital angular momentum then the g value for each spin state should have the free-electron value of 2·00. However, Bleaney and Bowers [8] have shown that the crystal field and spin−orbit coupling introduce an orbital contribution into the g value, which now becomes anisotropic, as discussed for axially distorted octahedral copper(II). Indeed, although within the present approximation the exchange interaction is considered to be isotropic, the magnetic susceptibility is anisotropic because of the anisotropy in the g value (the value of $\bar{g}$ used in Equation 7.10 is $\bar{g} = \frac{1}{3}(g_{\parallel} + 2g_{\perp})$). The temperature independent paramagnetism also arises by the mechanism discussed for mononuclear copper(II) complexes. More recently a modified interpretation of the magnetic susceptibility of $[Cu(acetate)_2 H_2 O]_2$ has been given [9]. In this treatment allowance is made for the possibility of pairing the electrons in one orbital, which introduces two extra spin-singlet levels into the energy diagram (Fig. 7.7). These two extra singlets must be included in the calculation of the magnetic susceptibility as they both may well be thermally populated at room temperature or above. The equation for the susceptibility now becomes:

$$\bar{\chi}_{Cu} = \frac{N\beta^2 6\bar{g}^2}{3kTF} + N\alpha, \qquad (7.15)$$

where $F = 3 + \exp(-J'/kT) + 2\exp(-J'/4kT) \cdot \cosh(2\gamma/kT)$, J' is the spin exchange integral and γ is a covalency parameter. It is claimed [9] that this model gives a better account of the experimental susceptibility data than the

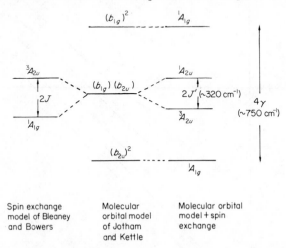

Figure 7.7. A comparison of energy levels produced by the Bleaney and Bowers, and Jotham and Kettle models for the exchange in copper(II) acetate monohydrate. (After Jotham and Kettle [9].)

simple model given by Equation 7.10, but this has been disputed [10]. However, if this model is correct then it implies that the exchange interaction is ferromagnetic in nature although the overall behaviour appears to be antiferromagnetic, because one is almost entirely observing thermal depopulation from the $^3A_{2u}$ into the $^1A_{1g}$ level as the temperature is lowered.

Binuclear chromium(III) complexes

Because octahedral mononuclear chromium(III) has an orbital singlet ground state, when it occurs in a binuclear complex in which each chromium has essentially this symmetry, the simple 'Dipolar Coupling' approach to the exchange interaction should be applicable. In this case $S = \frac{3}{2}$, and the possible spin states of the molecule are $S' = 3, 2, 1, 0$. Under these conditions Equation 7.5 leads to the following expression for the susceptibility:

$$\bar{\chi}_A = \frac{N\beta^2 \bar{g}^2}{3kT} \left[\frac{42 + 15\exp(6x) + 3\exp(10x)}{7 + 5\exp(6x) + 3\exp(10x) + \exp(12x)} \right], \qquad (7.16)$$

where $x = -(J/kT)$.

For reasons similar to those outlined in the case of copper(II) acetate monohydrate it may be necessary to use a value of $\bar{g}$ slightly different from the spin-only value and also a T.I.P. term. The variation of the magnetic susceptibility with temperature predicted by Equation 7.16 for various values of J is shown in Figs. 7.8 and 7.9, when $\bar{g} = 2$ and $N\alpha = 0$.

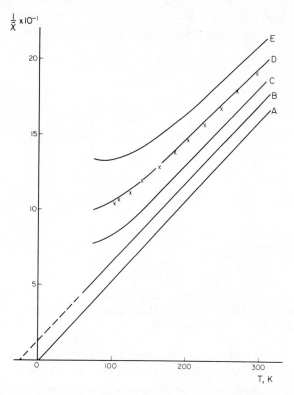

Figure 7.8. The variation of $1/\bar{\chi}_A$ against temperature for binuclear Cr(III) complexes. ×, experimental for $[(NH_3)_5 Cr . OH . Cr(NH_3)_4 H_2O] Br_5$. ─────, calculated from Equation 7.16 with $\bar{g} = 2 \cdot 0$ and A, $J = 0$; B, $J = -5$ cm^{-1}; C, $J = -10$ cm^{-1}; D, $J = -15$ cm^{-1}; E, $J = -20$ cm^{-1}.

An increase in the magnitude of J causes an increasing divergence from Curie-law behaviour. For $0 < J \leqslant -5$ cm^{-1} a Curie–Weiss law holds over the temperature range 60–300 K, θ increasing with increasing $|J|$. Further increases in $|J|$ cause deviation even from the Curie–Weiss law in the above temperature range, the deviation appearing as significant curvature in the $1/\chi$ *vs.* T plot. In fact when $J = -20$ cm^{-1} a maximum in the susceptibility occurs at ~90 K. Increasing the magnitude of $|J|$ causes the maximum in the susceptibility to move to higher temperatures and by the time $J = -100$ cm^{-1} the maximum occurs at ~265 K, see Fig. 7.9. Below this temperature the susceptibility steadily falls due to the thermal depopulation of the higher spin states until eventually the system becomes diamagnetic when only the $S' = 0$ state is thermally populated.

The application of Equation 7.16 to binuclear chromium(III) complexes may be illustrated with reference to the compounds $[(NH_3)_5Cr.OH.Cr(NH_3)_4-(H_2O)]Br_5$ and $[(NH_3)_5Cr.O.Cr(NH_3)_5]Br_4$ [11]. A plot of $1/\bar{\chi}_{Cr}$ vs. T for the

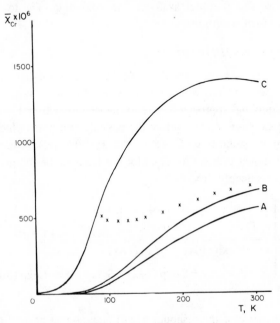

Figure 7.9. The variation of $\bar{\chi}_{Cr}$ with temperature for binuclear Cr(III) complexes. x, experimental for $[(NH_3)_5Cr.O.Cr(NH_3)_5]Br_4$. ————, calculated from Equation 7.16 with $\bar{g} = 2.0$ and A, $J = -200$ cm^{-1}; B, $J = -180$ cm^{-1}; C, $J = -100$ cm^{-1}.

first compound is shown in Fig. 7.8, which suggests that J is about -15 cm^{-1}. A more detailed interpretation of the data showed that $J = -14.4$ cm^{-1}, $\bar{g} = 1.99$ and $N\alpha = 56 \times 10^{-6}$ c.g.s. units gives a better interpretation of the data [11]. The variation of the magnetic susceptibility of the single oxo-bridged complex, $[(NH_3)_5Cr.O.Cr(NH_3)_5]Br_4$, is shown in Fig. 7.9. Owing to the difficulties in preparing pure samples of the compound the magnetic susceptibility does not fall rapidly enough with temperature (in fact a slight increase was observed below 110 K) compared with the theoretical predictions. However, it seems possible that if a pure specimen were obtained $-J$ would be of the order of 180 cm^{-1} or greater.

The large difference in the values of J for these two systems has been attributed to the different bridging bond angles and π-bonding abilities of the bridging groups [11]. In the oxo-bridged compound the Cr–O–Cr angle was assumed to be linear the oxygen using sp hybrid orbitals to form the σ-bonds. This then leaves two oxygen p orbitals to form a π-bonding system with the metal d_{xz} and d_{yz} orbitals. Such a π-bonding system for transmitting the exchange interaction in the hydroxo system seems to be less likely since the bond angle is probably closer to 120° and it would be expected that there would only be one oxygen p-orbital available for π-bonding. The lower value of J observed is compatible with this view.

Binuclear spin-free iron(III) complexes

Spin-free iron(III) has an orbital singlet ground state irrespective of the symmetry of the complex, which should make it an ideal system for applying the 'Dipolar Coupling' approach to any exchange interaction. For each iron(III) ion in a complex $S = \frac{5}{2}$, and therefore the possible spin states which arise due to the interaction in the dimer are $S' = 5, 4, 3, 2, 1, 0$. The relative energies of these spin states are shown in Fig. 7.6. Equation 7.5 leads to the following expression for the magnetic susceptibility:

$$\bar{\chi}_A = \frac{3N\beta^2\bar{g}^2}{3kT} \left[\frac{\begin{array}{l}55 + 30\exp(10x) + 14\exp(18x) \\ \qquad\qquad\qquad\qquad + 5\exp(24x) + \exp(28x)\end{array}}{\begin{array}{l}11 + 9\exp(10x) + 7\exp(18x) \\ \qquad\qquad + 5\exp(24x) + 3\exp(28x) + \exp(30x)\end{array}} \right], \qquad (7.17)$$

where $x = -J/kT$.

In this particular case there should be virtually no mixing in of excited states by spin–orbit coupling or magnetic field and thus we would expect $\bar{g}$ to be close to 2·00 and $N\alpha$ to be zero.

The variation of the magnetic susceptibility with temperature given by Equation 7.17 for a range of values of J is shown in Fig. 7.10. If there were no interaction between the iron atoms in the complex then the magnetic susceptibility would obey a Curie law with $\bar{\mu}$ equal to 5·92 B.M. Indeed, when calculating susceptibilities in interacting systems it is well worth putting $J = 0$ in any expression obtained in order to ensure that the susceptibility tends towards that expected for a mononuclear system as the exchange interaction disappears. The inclusion of a relatively small value of J, can cause considerable deviation from Curie-law behaviour, e.g. When $J = -2$ cm^{-1} $\bar{\mu}$ at 300 K is reduced to 5·7 B.M. and a Curie-Weiss law with $\theta = 20°$ is obeyed. As J is increased in magnitude the magnetic moment at 300 K decreases and θ in the Curie-Weiss law increases until eventually deviations from this law occur. These calculations show that in this particular system the variation of the magnetic susceptibility with temperature is very sensitive to small changes in J. In contrast to this in the

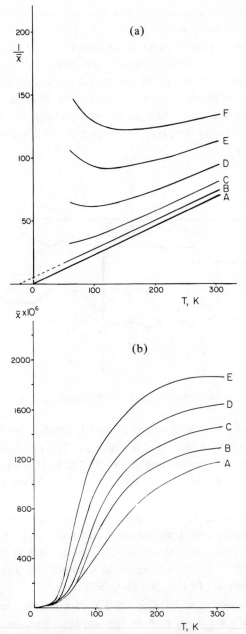

Figure 7.10. The calculated variation of susceptibility with temperature for binuclear iron(III) complexes with $\bar{g} = 2{\cdot}0$. (a) With A, $J = 0$; B, $J = -2$ cm^{-1}; C, $J = -5$ cm^{-1}; D, $J = -10$ cm^{-1}; E, $J = -15$ cm^{-1}; F, $J = -20$ cm^{-1}. (b) With A, $J = -120$ cm^{-1}; B, $J = -110$ cm^{-1}; C, $J = -100$ cm^{-1}; D, $J = -90$ cm^{-1}; E, $J = -80$ cm^{-1}.

$S = \frac{1}{2}$ binuclear system, J values of a few wave-numbers have little effect on the variation of the magnetic susceptibility with temperature, except at very low temperatures (less than about 10 K). The sensitivity of the susceptibility of the $S = \frac{5}{2}$ binuclear system to small changes in the value of J arises because of the relatively large spread in energy of the spin states of the molecule (the total spread is $30J$, i.e. 300 cm^{-1} when $|J| = 10$ cm^{-1}). This means that the higher spin states, which contribute most to the susceptibility, are thermally depopulated at relatively higher temperatures, leading to considerable deviations from the ideal Curie law.

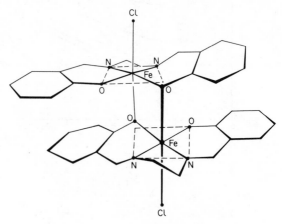

Figure 7.11. A diagram of the structure of (FesalenCl)$_2$.

An example of a binuclear spin-free iron(III) complex is (FesalenCl)$_2^*$, [12] the structure of which is illustrated in Fig. 7.11. The variation of the magnetic susceptibility over the temperature range 20–300 K [13] can be satisfactorily accounted for using Equation 7.16 with $J = -7\cdot5$ cm^{-1} and $\bar{g} = 2\cdot02$ (see Fig. 7.12).

With higher values of J, for example $J = -100$ cm^{-1}, the variation of the magnetic susceptibility with temperature is shown in Fig. 7.10(b). The susceptibility falls continuously with temperature, from its value at 300 K, eventually becoming zero at 0 K. The maximum in the susceptibility is very broad and for this order of magnitude of J it occurs above 300 K.

The magnetic moment at 300 K is 1·9 B.M.; for a system in which formally we considered each ion atom to have five unpaired electrons. This behaviour arises because the increased energy separations between the possible spin states of the molecule only allow the lower members to be thermally populated at the temperatures used. Thus the contribution to the susceptibility from the most

---

* salen = NN'-Ethylenebis-(salicylideneiminato).

paramagnetic spin states is lost (at 300 K with J of the order of -100 cm^{-1} states with $S' > 3$ give negligible contribution to the susceptibility). The effect on the magnetic susceptibility of ignoring the higher spin states is illustrated in Fig. 7.13.

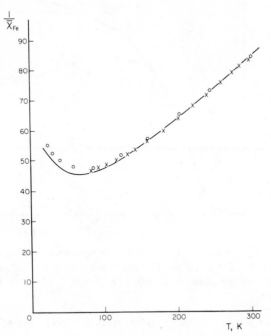

Figure 7.12. The variation of $1/\overline{\chi}_{Fe}$ against T for (FesalenCl)$_2$. ×, Experimental from Gerloch, Lewis, Mabbs, and Richards [13]. ○, Experimental from Rief, Long, and Baker [13]. ————, Calculated from Equation 7.17 with $\overline{g} = 2 \cdot 02$ and $J = -7 \cdot 5$ cm^{-1}.

When the temperature is lowered the number of molecules in the higher spin states rapidly decreases and hence the susceptibility falls, eventually becoming zero at 0 K when all the molecules are in the diamagnetic ground state.

Some examples of binuclear iron(III) complexes, the magnetic behaviour of which indicate J values of about -100 cm^{-1}, are provided by (Fesalen)$_2$O; (Fesalen)$_2$O . CH$_2$Cl$_2$; (Fesalen)$_2$O(pyridine)$_2$. The first of these compounds has been shown to be dimeric in solution and in the gas phase [14], whilst the last two have been shown by X-ray measurements to be oxo-bridged dimers with an FeOFe angle of ~140° [15]. The magnetic susceptibilities of all three compounds have been measured over the temperature range 80–300 K [14], and that of (Fesalen)$_2$O down to 20 K [13], and shown in Fig. 7.14. The increase in the magnetic susceptibility of (Fesalen)$_2$O below 25 K is almost certainly due to the presence of very small traces of mononuclear impurities. The magnetic

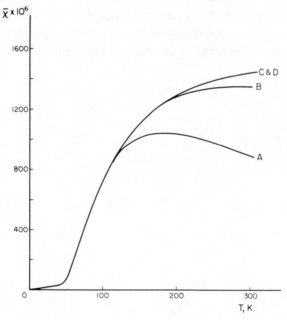

Figure 7.13. The variation of $\bar{\chi}_{Fe}$ with temperature in binuclear iron(III) complexes with $\bar{g} = 2 \cdot 0$ and $J = -100$ cm^{-1} when only the following spin states of the molecule are included: A, $S' = 0, 1$; B, $S' = 0, 1, 2$; C, $S' = 0, 1, 2, 3$; D, $S' = 0, 1, 2, 3, 4, 5$.

susceptibility of a mononuclear spin-free iron(III) compound at 20 K is of the order of $200\,000 \times 10^{-6}$ c.g.s. units and thus only $0 \cdot 1\%$ of this type of impurity would be needed to account for the experimental data.

As in the case of the chromium(III) compounds discussed previously, the type of bridging group and its geometry play an important part in determining the value of J required to account for the experimental data. Again we see that a single oxo-bridge, even though it gives a non-linear arrangement, is more efficient at transmitting the exchange interaction than an oxygen atom which forms three bonds.

7.3.2. TRIANGULAR TRINUCLEAR COMPLEXES

If all the interacting ions in the complex are of the same chemical type and are at the corners of an equilateral triangle they will all be equivalent. In this case the Hamiltonian describing the interaction is:

$$\mathcal{H} = -2J(\hat{S}_1 . \hat{S}_2 + \hat{S}_2 . \hat{S}_3 + \hat{S}_1 . \hat{S}_3). \tag{7.18}$$

Since all the interacting ions are equivalent, the possible spin states of the molecule will be:

$$S' = 3S, 3S - 1, 3S - 2, \ldots 0 \text{ or } \tfrac{1}{2}$$

(depending on whether S is integral or half integral).

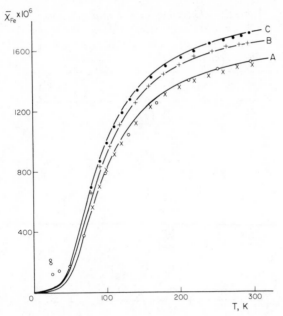

Figure 7.14. The variation of $\bar{\chi}_{Fe}$ with T for: A. (Fesalen)$_2$O; ×, experimental from ref. 14; ○, experimental from Rief, Long, and Baker [13]; ———, calculated from Equation 7.17 with $\bar{g}$ = 2·0 and J = −95 cm^{-1}. B. (Fesalen)$_2$O-H pyridine. +, experimental; ———, calculated from Equation 7.17 with $\bar{g}$ = 2·0 and J = −90 cm^{-1}. C. (Fesalen)$_2$OCH$_2$Cl$_2$ ●, experimental, ———, calculated from Equation 7.17 with $\bar{g}$ = 2·0 and J = −87 cm^{-1}.

The Equations 7.2 to 7.5, with z = 2 and n = 3, are used to calculate the magnetic susceptibility. Kambe's extension of the above treatment to the case when all three atoms are not equivalent will also be outlined in this section.

(a) Chromium(III) complexes

Each chromium(III) ion in such a complex has formally $S = \frac{3}{2}$, which leads to the following possible spin states of the molecule:

$$S' = \tfrac{9}{2}, \tfrac{7}{2}, \tfrac{5}{2}, \tfrac{3}{2}, \tfrac{1}{2}.$$

The energies (neglecting the constant term $3S(S + 1)$) and the number of times each state can occur are summarized in Table 7.1. The expression for the magnetic susceptibility is:

$$\bar{\chi}_A = \frac{N\beta^2\bar{g}^2}{3kT}\frac{1}{4}\left[\frac{\begin{array}{l}165 \exp(-24x) + 168 \exp(-15x) \\ \quad + 105 \exp(-8x) + 40 \exp(-3x) + 2\end{array}}{\begin{array}{l}5 \exp(-24x) + 8 \exp(-15x) \\ \quad + 9 \exp(-8x) + 8 \exp(-3x) + 2\end{array}}\right]$$

$$(7.19)$$

where $x = -J/kT$.

Table 7.1. *The energies and multiplicities of the various spin-states. for a triangular arrangement of octahedrally co-ordinated chromium(III) ions.*

S'	$-W(S')$	$\Omega(S')$
$\dfrac{9}{2}$	$\dfrac{9 \times 11J}{4}$	1
$\dfrac{7}{2}$	$\dfrac{7 \times 9J}{4}$	2
$\dfrac{5}{2}$	$\dfrac{5 \times 7J}{4}$	3
$\dfrac{3}{2}$	$\dfrac{3 \times 5J}{4}$	4
$\dfrac{1}{2}$	$\dfrac{3J}{4}$	2

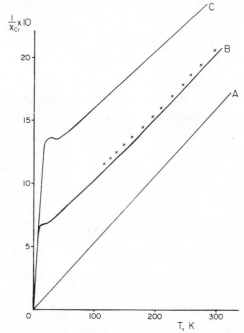

Figure 7.15. The variation of $1/\overline{\chi}_{Cr}$ against T for an equilateral triangle of Cr(III) ions. x, experimental for $[Cr_3O(acetate)_6(H_2O)_3]Cl6H_2O$ ————, calculated from Equation 7.19 with $\overline{g} = 2 \cdot 0$ and Å, $J = 0$; B, $J = -10$ cm^{-1}; C, $J = -20$ cm^{-1}.

In the limit of large values of x (i.e. either low temperature or large values of $-J$) this model predicts that the ground state of the *molecule* has one unpaired electron. This means that $\bar{\mu}$ per chromium ion tends to 1 B.M. under these conditions. The behaviour of $1/\bar{\chi}_A$ vs. T, for a selected range of J, is illustrated in Fig. 7.15.

If the triangular arrangement deviates from that of an equilateral triangle then Kambe's approach must be used, since all the interacting ions are no longer equivalent. When the arrangement of atoms in the complex is that of an isosceles triangle, with atom number two at the apex, the Hamiltonian for the interaction is:

$$\mathcal{H} = -2J[\hat{S}_1 . \hat{S}_2 + \hat{S}_2 . \hat{S}_3 + \alpha \hat{S}_1 . \hat{S}_3] \qquad (7.20)$$

where $J_{12} = J_{23} = J$ and $J_{13} = \alpha J$.

The energies which arise from this Hamiltonian are most readily obtained by expressing the operators $\hat{S}_i . \hat{S}_j$ in terms of spin quantum numbers for the *whole system* and the spin quantum numbers of the individual paramagnetic ions. If we write new quantum numbers for the complex as

$$S^* = S_1 + S_3 \qquad \text{and} \qquad S' = S_1 + S_2 + S_3$$

we can then have

$$(\hat{S}^*)^2 = (\hat{S}_1 + \hat{S}_3)^2 = \hat{S}_1^2 + \hat{S}_3^2 + 2\hat{S}_1 . \hat{S}_3,$$

or

$$2\hat{S}_1 . \hat{S}_3 = (\hat{S}^*)^2 - \hat{S}_1^2 - \hat{S}_3^2. \qquad (7.21)$$

Since $\hat{S}^2 |S\rangle = S(S+1) |S\rangle$ and taking $S_1 = S_2 = S_3$, Equation 7.21 becomes:

$$2\hat{S}_1 . \hat{S}_3 = S^*(S^* + 1) - 2S(S + 1). \qquad (7.22)$$

Similarly $(\hat{S}')^2 = (\hat{S}_1 + \hat{S}_2 + \hat{S}_3)^2$ gives:

$$2\hat{S}_1 . \hat{S}_2 + 2\hat{S}_2 . \hat{S}_3 = S'(S' + 1) + S^*(S^* + 1) - S(S + 1). \qquad (7.23)$$

When these expressions for the spin operators are substituted into the Hamiltonian we have:

$$\mathcal{H} = -J[S'(S' + 1) - (1 - \alpha)S^*(S^* + 1) - (1 + 2\alpha)S(S + 1)] \qquad (7.24)$$

and the energy of each state specified by S' is:

$$W(S') = -J[S'(S' + 1) - (1 - \alpha)S^*(S^* + 1) - (1 + 2\alpha)S(S + 1)]. \qquad (7.25)$$

The possible values of S^* and S' are obtained as follows:

$$S^* = S_1 + S_3, S_1 + S_3 - 1, S_1 + S_3 - 2, \dots . |S_1 - S_3|$$
$$= 2S, 2S - 1, 2S - 2, \dots 0$$

if the ions are identical. For each of these possible values of $S*$ we have

$$S' = S* + S_2, S* + S_2 - 1, S* + S_2 - 2, \ldots |S* - S_2|$$
$$= S* + S, S* + S - 1, \ldots |S* - S|$$

if the ions are identical.

The magnetic susceptibility can now be obtained by substituting this information into Van Vleck's susceptibility equation. It is important to note that some states with the same value of S' may have different energies and care must be taken when performing the summation over all states in calculating the susceptibility. The treatment outlined here reduces to that given for the equilateral triangle if $\alpha = 1$.

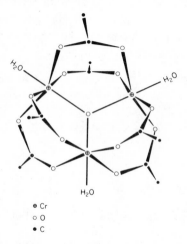

Figure 7.16. A diagram of the structure of the $[Cr_3 O(acetate)_6 (H_2 O)_3]^+$ ion. (After Figgis and Robertson [16].)

The treatments outlined here have been applied to the compound $[Cr_3O(acetate)_6(H_2O)_3]Cl . 6H_2O$, the structure of which contains a central oxygen atom surrounded by a triangular arrangement of chromium atoms [16], see Fig. 7.16. The susceptibility data between 100–300 K (see Fig. 7.15) may be satisfactorily accounted for with $\bar{g} = 2 \cdot 00$ and $J = -10 \cdot 4$ cm^{-1} and $\alpha = 1$ [17]. However, results at lower temperatures seem to be better interpreted with $J = -10 \cdot 4$ cm^{-1} and $\alpha = 1 \cdot 25$ [18]. At temperatures between 0·4 and 4·2 K the susceptibility follows the equation $\bar{\chi}_{Cr} = 0 \cdot 131/(T + 0 \cdot 13)$ [19], compared with $\bar{\chi}_{Cr} = 0 \cdot 125/T$ predicted by Equation 7.19. The reasons for the discrepancy between the theoretical predictions and the observed behaviour is at present unknown but it is possible that the Curie–Weiss law behaviour at temperatures below 4·2 K arises from intermolecular exchange interactions.

(b) Spin-free iron(III) complexes

In these complexes each iron(III) ion is formally described by $S = \frac{5}{2}$ and the possible spin states of the molecule which arise for an equilateral triangular arrangement are:

$$S' = 15/2, 13/2, 11/2, 9/2, 7/2, 5/2, 3/2, 1/2.$$

The energies (neglecting the constant term $3S(S + 1)$), and the number of times each of these spin states can occur, are given in Table 7.2. The substitution of this information into Equation 7.5 leads to the following expression for the susceptibility:

$$\bar{\chi_A} = \frac{N\beta^2 \bar{g}^2}{3kT} \frac{1}{4} \left[\frac{340 \exp(-63x) + 455 \exp(-48x)}{4 \exp(-63x) + 7 \exp(-48x)} \right.$$

$$+ \frac{429 \exp(-35x) + 330 \exp(-24x) + 210 \exp(-15x)}{+ 9 \exp(-35x) + 10 \exp(-24x) + 10 \exp(-15x)}$$

$$\left. + \frac{105 \exp(-8x) + 20 \exp(-3x) + 1}{+ 9 \exp(-8x) + 4 \exp(-3x) + 1} \right], \quad (7.26)$$

where $x = -J/kT$.

Table 7.2. *The energies and multiplicities of the various spin-states for a triangular arrangement of spin-free iron(III) ions.*

S'	$-W(S')$	$\Omega(S')$
$\dfrac{15}{2}$	$\dfrac{15 \times 17J}{4}$	1
$\dfrac{13}{2}$	$\dfrac{13 \times 15J}{4}$	2
$\dfrac{11}{2}$	$\dfrac{11 \times 13J}{4}$	3
$\dfrac{9}{2}$	$\dfrac{9 \times 11J}{4}$	4
$\dfrac{7}{2}$	$\dfrac{7 \times 9J}{4}$	5
$\dfrac{5}{2}$	$\dfrac{5 \times 7J}{4}$	6
$\dfrac{3}{2}$	$\dfrac{3 \times 5J}{4}$	4
$\dfrac{1}{2}$	$\dfrac{3J}{4}$	2

As in the chromium(III) system the model predicts one unpaired electron in the ground state of the *molecule* or $\bar{\mu} = 1$ B.M. per iron.

The magnetic susceptibility of $[Fe_3O(acetate)_6(H_2O)_3]Cl5H_2O$ in the range 90–300 K has been interpreted in terms of Equation 7.26 [17]. It was found that $\bar{g} = 2 \cdot 00$ and $J = -30 \cdot 3$ cm^{-1} gave a good fit to the observed behaviour, see Fig. 7.17(a). However, if the comparison is made on the basis of the experimental and calculated susceptibilities, Fig. 7.17(b), which is a much more sensitive test, very significant deviations between the experimental and theoretical curves appear below about 160 K. This feature serves as a warning that care should be taken when comparing experimental and theoretical curves in that, when using $\bar{\mu}$, significant variations of susceptibility with temperature may well be smoothed out. The reasons for these differences are not yet known, but it seems possible that either the compound does not have the same structure as the chromium analogue, although they have been reported to be isomorphous [17], or the sample was insufficiently pure, or the theory is inadequate.

7.3.3. LINEAR TRINUCLEAR COMPLEXES

In this type of compound the interacting ions are not all equivalent and Kambe's method is used to determine the energies of the spin-states of the molecule. If we assume that only the interactions between nearest neighbours are important then the Hamiltonian representing the interaction can be written as:

$$\mathscr{H} = -2J[\hat{S}_1 . \hat{S}_2 + \hat{S}_2 . \hat{S}_3]. \tag{7.27}$$

Defining new quantum numbers for the system as $S^* = S_1 + S_3$ and $S' = S_1 + S_2 + S_3$ we can write

$$2\hat{S}_1 . \hat{S}_2 + 2\hat{S}_2 . \hat{S}_3 = S'(S' + 1) - S^*(S^* + 1) - S(S + 1)$$

and

$$W(S') = -J[S'(S' + 1) - S^*(S^* + 1) - S(S + 1)]. \tag{7.28}$$

Assuming that the interacting ions are of the same spin type, the possible values of S^* and S'' are:

$$S^* = 2S, 2S - 1, \ldots 0,$$

and for each of these values of S^*,

$$S' = S^* + S, S^* + S - 1, \ldots |S^* - S|.$$

If we consider the simplest example where each interacting ion formally has one unpaired electron, i.e. $S = \frac{1}{2}$, the values of the spin quantum numbers and energies of the possible spin states of the molecule are summarized in Table 7.3.

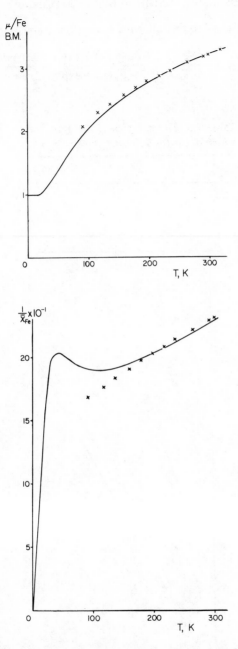

Figure 7.17. The magnetic behaviour with temperature of [Fe$_3$O(acetate)$_6$-(H$_2$O)$_3$]Cl5H$_2$O. (a) $\bar{\mu}_{Fe}$, x experimental, (b) 1/$\bar{\chi}_{Fe}$, x, experimental. In both (a) and (b) the solid curve is that calculated from Equation 7.26 with $\bar{g}$ = 2·0 and J = − 30·3 cm^{-1}.

These energy levels lead to the following susceptibility equation:

$$\bar{\chi}_M = \frac{\bar{g}^2 N \beta^2}{3kT} \frac{3}{4} \left[\frac{10 \exp(x) + \exp(-2x) + 1}{2 \exp(x) + \exp(-2x) + 1} \right], \tag{7.29}$$

where $x = J/kT$.

Table 7.3. *The energies of the spin states which occur for a linear trimer with $S = \frac{1}{2}$.*

S'	S^*	$-W(S')$
$\frac{3}{2}$	1	J
$\frac{1}{2}$	1	$-2J$
$\frac{1}{2}$	0	0

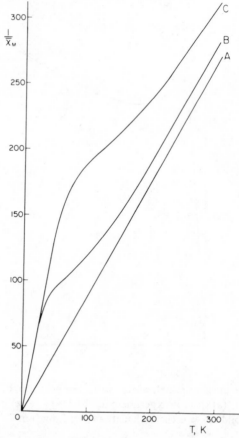

Figure 7.18. The variation of $1/\bar{\chi}_M$ against T for a linear trimer with $S = \frac{1}{2}$. Curves calculated from Equation 7.29 with $\bar{g} = 2\cdot0$ and A, $J = 0$; B, $J = -100$ cm^{-1}; C, $J = -200$ cm^{-1}.

The behaviour of the magnetic susceptibility against temperature, as expressed by Equation 7.29, for a range of values of J, with $\bar{g} = 2\cdot00$ is illustrated in Fig. 7.18. The effect of increasing the magnitude of J is relatively small at 300 K, $\bar{\mu}$ per atom being 1·70 B.M. for $J = -100$ cm^{-1}, 1·61 B.M. for $J = -200$ cm^{-1} compared with 1·73 B.M. for $J = 0$. Thus measurements at room temperature would not be very good for detecting the presence of exchange interactions in such a system unless J is very large in magnitude. It is interesting to note that in this case there are two possible spin states with $S' = \frac{1}{2}$, but they have different energies. This is in contrast to the simpler approach of assuming each ion to be identical and taking an average value for the number of nearest neighbours. In this latter case there are still two levels with $S' = \frac{1}{2}$, but they are now degenerate.

The use of the alternative method of calculation given in Section 7.3.1 provides a useful insight into the origin of the differences in energy of the two $S = \frac{1}{2}$ spin states. The wave-functions may be written in terms of $|m_{s1}, m_{s2}, m_{s3}\rangle$, which for three ions each with $S = \frac{1}{2}$ gives:

$$\psi_1 = |\tfrac{1}{2}, \tfrac{1}{2}, \tfrac{1}{2}\rangle \qquad\qquad \psi_6 = |-\tfrac{1}{2}, -\tfrac{1}{2}, \tfrac{1}{2}\rangle$$
$$\psi_2 = |-\tfrac{1}{2}, -\tfrac{1}{2}, -\tfrac{1}{2}\rangle \qquad \psi_7 = |-\tfrac{1}{2}, \tfrac{1}{2}, -\tfrac{1}{2}\rangle$$
$$\psi_3 = |-\tfrac{1}{2}, \tfrac{1}{2}, \tfrac{1}{2}\rangle \qquad\quad\; \psi_8 = |\tfrac{1}{2}, -\tfrac{1}{2}, -\tfrac{1}{2}\rangle$$
$$\psi_4 = |\tfrac{1}{2}, -\tfrac{1}{2}, \tfrac{1}{2}\rangle$$
$$\psi_5 = |\tfrac{1}{2}, \tfrac{1}{2}, -\tfrac{1}{2}\rangle$$

The secular determinant resulting from the application of the spin Hamiltonian (7.3.27) to the above functions is:

	$|\psi_1\rangle$	$|\psi_2\rangle$	$|\psi_3\rangle$	$|\psi_4\rangle$	$|\psi_5\rangle$	$|\psi_6\rangle$	$|\psi_7\rangle$	$|\psi_8\rangle$
$\langle\psi_1|$	$-J-E$	0	0	0	0	0	0	0
$\langle\psi_2|$	0	$-J-E$	0	0	0	0	0	0
$\langle\psi_3|$	0	0	$0-E$	$-J$	0	0	0	0
$\langle\psi_4|$	0	0	$-J$	$J-E$	$-J$	0	0	0
$\langle\psi_5|$	0	0	0	$-J$	$0-E$	0	0	0
$\langle\psi_6|$	0	0	0	0	0	$0-E$	$-J$	0
$\langle\psi_7|$	0	0	0	0	0	$-J$	$J-E$	$-J$
$\langle\psi_8|$	0	0	0	0	0	0	$-J$	$0-E$

$$= 0. \quad (7.30)$$

The solution to the secular determinant (7.30) gives:

$$E = -J \quad \text{four times}$$
$$E = 2J \quad \text{twice}$$
$$E = 0 \quad \text{twice}$$

which is precisely the result summarized in Table 7.3. It is now interesting to find the wave-functions corresponding to these energies and these are:

At energy −J

$$\phi_1 = \psi_1$$
$$\phi_2 = \psi_2$$
$$\phi_3 = \sqrt{\tfrac{1}{3}}(\psi_4 + \psi_3 + \psi_5)$$
$$\phi_4 = \sqrt{\tfrac{1}{3}}(\psi_7 + \psi_6 + \psi_8)$$

At energy 2J

$$\phi_5 = \sqrt{\tfrac{1}{6}}(2\psi_4 - \psi_3 - \psi_5)$$
$$\phi_6 = \sqrt{\tfrac{1}{6}}(2\psi_7 - \psi_6 - \psi_8)$$

At energy 0

$$\phi_7 = \sqrt{\tfrac{1}{2}}(\psi_3 - \psi_5)$$
$$\phi_8 = \sqrt{\tfrac{1}{2}}(\psi_6 - \psi_8)$$

A comparison between the functions ϕ_i and the spin states represented by S' in Table 7.3 may now be made. It can readily be seen that the functions $\phi_1, \phi_2,$ ϕ_3, ϕ_4 belong to a state total spin of $\tfrac{3}{2}$ since for ϕ_1, $\Sigma_{i=1}^{3} m_{si} = \tfrac{3}{2}$ and for ϕ_4, $\Sigma_{i=1}^{3} m_{si} = \tfrac{1}{2}$. The functions ϕ_2 and ϕ_3 contain functions which have $\Sigma_{i=1}^{3} m_{si} = \tfrac{1}{2}$ and $-\tfrac{1}{2}$ respectively. The functions ϕ_5 and ϕ_6, and likewise ϕ_7 and ϕ_8 belong to states with total spin $\tfrac{1}{2}$ since they each have $\Sigma_{i=1}^{3} m_{si} = \tfrac{1}{2}$ and $-\tfrac{1}{2}$ respectively. The different energies of the $S' = \tfrac{1}{2}$ levels arise because of differing ways in which the spins on the individual centres combine.

There seems to be only one well established example of a linear trimer, namely Ni(acetylacetonato)$_2$ in which each nickel atom is bridged by an oxygen atom of the acetylacetonate group. The magnetic susceptibility of this compound has been studied in the temperature range 300 to 0·365 K [20]. The interpretation of the results is more complicated than for the simple systems that we have studied so far, in that the nearest neighbour exchange is ferromagnetic and there is also an antiferromagnetic interaction between the terminal atoms in the chain. Under these conditions the Hamiltonian for the interaction is

$$\mathscr{H} = -2J[\hat{S}_1 . \hat{S}_2 + \hat{S}_2 . \hat{S}_3] - 2J_{31}\hat{S}_3 . \hat{S}_1. \tag{7.31}$$

The spin states and their energies arising from the interaction are summarized in Table 7.4 and the expression for the magnetic susceptibility is:

$$\overline{\chi}_{Ni} = \frac{K}{T}\,\frac{2}{3}\left[\frac{42\exp[2(2x+y)] + 15\exp[2(y-x)] + 15\exp[2(x-y)]}{7\exp[2(2x+y)] + 5\exp[2(y-x)] + 5\exp[2(x-y)]}\right.$$

$$\left.\frac{+3\exp[2(y-3x)]+3\exp[-2(x+y)]+3\exp[-4y]}{+3\exp[2(y-3x)]+3\exp[-2(x+y)]+3\exp[-4y]+\exp[-2(2x+y)]}\right],$$

(7.32)

where $x = J/kT$, $y = J_{31}/kT$, and $K = (\overline{g}^2 N\beta^2/3k)$.

After allowing for a T.I.P. contribution to the susceptibility of 230×10^{-6} c.g.s. the experimental data was accounted for using Equation 7.32 with $\overline{g} = 2.06$, $J = +26$ cm^{-1}, and $J_{31} = -7$ cm^{-1}, see Fig. 7.19.

Table 7.4. *The energies of the spin states which occur for* [Ni(acetylacetonato)$_2$]$_3$.

S'	S^*	$-W(S')$
3	2	$+4J + 2J_{31}$
2	2	$-2J + 2J_{31}$
2	1	$+2J - 2J_{31}$
1	2	$-6J + 2J_{31}$
1	1	$-2J - 2J_{31}$
1	0	$-4J_{31}$
0	1	$-4J - 2J_{31}$

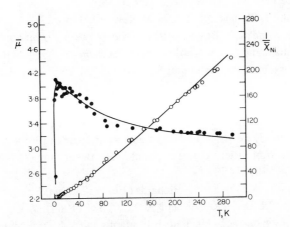

Figure 7.19. The temperature variation of $\overline{\mu}$ per Ni atom and $1/\overline{\chi}_{Ni}$ for [Ni(acetylacetonato)$_2$]$_3$. ○, experimental $1/\overline{\chi}_{Ni}$; ●, experimental $\overline{\mu}$ calculated from $\overline{\mu} = 2.8273\,[(\overline{\chi}_{Ni} - N\alpha)T]^{½}$ with $N\alpha = 230 \times 10^{-6}$. ———, calculated from Equation 7.32 with $\overline{g} = 2.06$, $J = +26$ cm^{-1}, $J_{31} = -7$ cm^{-1}. (Reproduced with permission from reference 20.)

Concluding remarks

In this chapter we have attempted to show the reader how principles established for mononuclear transition metal complexes may be carried over to polynuclear complexes, and how the subsequent interaction between the magnetic centres may be understood using molecular orbital theory. Further, we have attempted to illustrate how the 'Dipolar Coupling' approach may be applied to the calculation of the magnetic susceptibilities of some polymeric arrangements of metal atoms. This approach has been compared where possible with results on systems of known geometry. Inevitably geometries not covered in the text are known and further theoretical expressions for some of these may be found in references 21 and 22, although an understanding of the material contained within this chapter should enable the reader to obtain the required equations.

REFERENCES

1. B.N. Figgis and R.L. Martin (1956). *J. Chem. Soc.*, 3837.
2. H.A. Kramers (1934). *Physica*, **1**, 182.
3. J.H.E. Griffiths, J. Owen and I.M. Ward (1953). *Proc. Roy. Soc. (London)*, **A219**, 526; K.W.H. Stevens (1953). *Proc. Roy. Soc. (London)*, **A219**, 542.
4. J.H. Van Vleck (1965). *Electric and Magnetic Susceptibilities*, Oxford University Press.
5. L. Dunitz and L.E. Orgel (1953). *J. Chem. Soc.*, 2594.
6. K. Kambe (1950). *J. Phys. Soc. Japan*, **5**, 48.
7. H. Abe and J. Shimada (1953). *Phys. Rev.*, **90**, 316.
8. B. Bleaney and K.D. Bowers (1952). *Proc. Roy. Soc. (London)*, **A214**, 451.
9. R.W. Jotham and S.F.A. Kettle (1969). *J. Chem. Soc. (A)*, 2816, 2821.
10. A.K. Gregson, R.L. Martin and S. Mitra (1971). *Proc. Roy. Soc. (London)*, **A320**, 473.
11. A. Earnshaw and J. Lewis (1961). *J. Chem. Soc.*, 396.
12. M. Gerloch and F.E. Mabbs (1967). *J. Chem. Soc. (A)*, 1900.
13. M. Gerloch, J. Lewis, F.E. Mabbs and A. Richards (1968). *J. Chem. Soc. (A)*, 112; W.M. Rief, G.J. Long and W.A. Baker, Jun. (1968). *J. Amer. Chem. Soc.*, **90**, 6347.
14. J. Lewis, F.E. Mabbs and A. Richards (1967). *J. Chem. Soc. (A)*, 1014.
15. M. Gerloch, E.D. McKenzie and A.D.C. Towl (1969). *J. Chem. Soc. (A)*, 2850; P. Coggon, A.T. McPhail, F.E. Mabbs and V.N. McLachlan (1971). *J. Chem. Soc. (A)*, 1014.
16. B.N. Figgis and G. Robertson (1965). *Nature*, **205**, 694.
17. A. Earnshaw, B.N. Figgis and J. Lewis (1966). *J. Chem. Soc. (A)*, 1656.
18. J. Wucher and H.M. Gijsman (1954). *Physica*, **20**, 361.
19. J.T. Schriempf and S.A. Friedberg (1964). *J. Chem. Phys.*, **40**, 296.

20. A.P. Ginsberg, R.L. Martin and R.C. Sherwood, (1968). *Inorg. Chem.*, 7, 932.
21. C.G. Barraclough, H.B. Gray and L. Dubicki (1968). *Inorg. Chem.*, 7, 844.
22. R.W. Jotham and S.F.A. Kettle (1970). *Inorg. Chim. Acta.*, 4, 145.

Index